AF544926

LOW-TEMPERATURE OXIDATION

THE CORROSION MONOGRAPH SERIES

K. Nobe, R. T. Foley, J. Kruger, H. Leidheiser, Jr., J. B. Wagner, Jr., Editors (1981–)

R. T. Foley, Z. A. Foroulis, N. Hackerman, C. V. King, F. L. La Que, H. H. Uhlig, Editors (1962–1981)

Ailor, *Atmospheric Corrosion*
Friend, *Corrosion of Nickel and Nickel Base Alloys*
Sedriks, *Corrosion of Stainless Steels*
Frankenthal and Kruger, *Passivity of Metals*
LaQue, *Marine Corrosion*
Berry, *Corrosion in Nuclear Applications*
Ailor, *Handbook on Corrosion Testing and Evaluation*
Kofstad, *High-Temperature Oxidation of Metals*
Logan, *The Stress Corrosion of Metals*
Godard, Jepson, Bothwell, and Kane, *The Corrosion of Light Metals*
Leidheiser, *The Corrosion of Copper, Tin, and Their Alloys*
Fehlner, *Low-Temperature Oxidation: The Role of Vitreous Oxides*

LOW-TEMPERATURE OXIDATION
The Role of Vitreous Oxides

FRANCIS P. FEHLNER
Research and Development Division
Corning Glass Works
Corning, New York

Sponsored by
THE ELECTROCHEMICAL SOCIETY, INC. *Pennington, New Jersey*

A Wiley-Interscience Publication

John Wiley & Sons

New York • Chichester • Brisbane • Toronto • Singapore

Copyright © 1986 by John Wiley & Sons, Inc.

All rights reserved. Published simultaneously in Canada.

Reproduction or translation of any part of this work beyond that permitted by Section 107 or 108 of the 1976 United States Copyright Act without the permission of the copyright owner is unlawful. Requests for permission or further information should be addressed to the Permissions Department, John Wiley & Sons, Inc.

Library of Congress Cataloging in Publication Data:

Fehlner, F. P.
Low-temperature oxidation.

(Corrosion monograph series, ISSN 0190-0994)
"Sponsored by the Electrochemical Society, Inc."
Includes index.
1. Oxidation. 2. Metallic oxides. I. Electrochemical Society. II. Title. III. Series.
QD501.F354 1986 541.3'93 85-22491
ISBN 0-471-87448-5

Printed in the United States of America

10 9 8 7 6 5 4 3 2 1

To Professor N. F. Mott,

who provided the inspiration for this work

PREFACE TO THE SERIES

The *Corrosion Monograph Series* is the mechanism chosen by the Corrosion Division of The Electrochemical Society to bring the corrosion literature up to date.

The field of corrosion has seen a tremendous amount of research and development stimulated by problems related to great advances in technology and facilitated by parallel advances in corrosion science. This has resulted in a corresponding expansion in the published literature with the appearance of numerous new concepts and a large body of new data in many areas. The preparation of monographs in selected areas, to be followed by appropriate revision, was chosen as the preferred course to keep pace with progress in the fields of corrosion science and corrosion engineering.

These monographs are being written by specialists in each field, and the number of authors contributing to a single volume has been intentionally kept small to provide for such sharpening of point of view as is offered by small areas, but not large ones. Also, this accelerates publication.

The nature of the presentation, that is, the relative emphasis on the scientific aspects as contrasted to the engineering aspects, varies with the subject dealt with, the nature of the pertinent data, and the response of the authors to these influences. The several volumes include critical exposition of modern theories as well as specific data of the handbook type so useful to the practicing engineer. With this approach, the Series is designed to be useful to basic scientists working in research as well as to engineers confronted with corrosion problems in the field.

The Corrosion Monograph Series Committee

K. Nobe, Chairman
R. T. Foley
J. Kruger
H. Leidheiser, Jr.
J. B. Wagner, Jr.

PREFACE

The concept of the present work had its genesis in 1969 during a short sabbatical with Professor N. F. Mott at the Cavendish. Our objective was to update the model of low-temperature oxidation as put forward by Cabrera and Mott in 1948–1949. In this work, oxide growth was explained in terms of a spontaneously generated electric field which lowered the activation energy for cation movement into an oxide. Subsequently, results of Davies, Domeij, Pringle, and Brown showed that both anions and cations can be mobile in anodic oxides. This observation was considered relevant to low-temperature oxidation as well as anodization since both processes are field driven.

One result of the 1969 work was a more complete description of low-temperature oxidation. Fehlner and Mott divided the process into three consecutive parts. A clean metal surface adsorbs one or more monolayers of oxygen to form a two-dimensional structure. This converts to a three-dimensional oxide through island growth by a mechanism such as place exchange. Place exchange results from the cooperative movement of two or more ions and is favored because there is no fixed oxide structure at this point in the process. Hence, it is difficult to describe island growth in terms of independent cation and/or anion movement through an oxide network or lattice. Oxidation kinetics change from fast linear to slow logarithmic with the completion of lateral island growth. It is at this point that the differentiation between network-forming (vitreous) and network modifying (polycrystalline) oxides can be observed most clearly. This distinction forms a major theme of the present book.

The chapters cover varying aspects of low-temperature oxidation, from adsorption to slow logarithmic growth. Each topic is treated comprehensively with respect to oxidation. However, the discussions should not be construed as an exhaustive study of each particular discipline. They are meant to serve as a guide to current theories and models with the goal of extending models of low-temperature oxidation beyond the limited knowledge of today. Where possible, conflicts in the literature are resolved, but when data are limited, both sides of a question are presented and the conclusion is left to the reader and further research.

Silicon oxidation is treated in detail since it is the foremost example of vitreous oxide formation during thermal oxidation. By combining low-temperature oxidation of metals with silicon oxidation, this book is particularly useful to those concerned with electronic devices. As such, it complements a companion volume in *The Corrosion Monograph Series,* one concerned with high-temperature oxidation and written by Per Kofstad.

Many people have contributed to making this book possible. First and foremost is Professor N. F. Mott, to whom the book is dedicated. His ability to elicit new ideas created the challenge and focus of the work. Professor R. T. Foley's continuing support while Chairman of The Corrosion Monograph Series Committee of The Electrochemical Society provided the guidance needed to write the book. It was completed under the Chairmanship of Professor K. Nobe.

Comments and contributions of reviewers were necessary and welcome. I am indebted to the following people, who read and in some cases helped revise portions of the book:

Professor J. M. Blakely
Dr. N. F. Borrelli
Dr. M. Cohen
Dr. B. E. Deal
Professor M. J. Dignam
Dr. D. E. Fowler
Dr. R. P. Frankenthal
Professor A. T. Fromhold, Jr.
Dr. E. Kooi
Professor N. F. Mott
Dr. A. G. Revesz
Professor M. W. Roberts
Professor H. A. Schaeffer
Professor W. A. Tiller

and to others known only to the Chairman of The Corrosion Monograph Series Committee.

Equally important was the silent support of my family, Mary, Paul, John, and Joe. I am particularly indebted to them for allowing me the quiet time needed for writing.

The help provided by Dr. T. P. Hoffman and the staff of John Wiley & Sons and the continued support of Corning Glass Works are greatly appreciated. I am especially indebted to Brenda Boller, who proofread the manuscript, and to the typists who carried out the many text revisions over the past 10 years, in particular Brenda Powell and Carol Stone.

FRANCIS P. FEHLNER

Corning, New York
January 1986

CONTENTS

LIST OF TABLES

SYMBOLS

A	Twice the characteristic oxide thickness
A'	Activation term
A''	Thermionic emission constant
A_n	Constants ($n = 1, 2, 3, \ldots$)
a	Half the distance separating two adjacent equilibrium ion sites in an oxide
a_d	Average length of a dipole
a_m	Number of "melting" events per atom
a_0	Distance advanced by a growing crystal face in unit kinetic process
a', a''	Constants
B	Parabolic rate constant
B/A	Linear rate constant
B_n, B', B''	Constants ($n = 1, 2, 3, \ldots$)
b, b'	Constants
C	Concentration
C_1	Negative nonbridging chalcogen
C_3	Positive threefold coordinated chalcogen
C_i'	Initial concentration
C_i	Oxidant concentration on the inner oxide surface
C_o	Oxidant concentration on the outer oxide surface
C^*	Equilibrium oxidant concentration at the oxide surface
c, c', c''	Constants
D	Diffusion coefficient

D_{eff}	Effective diffusion coefficient
D^-	Negative nonbridging chalcogen
D^+	Positive threefold coordinated chalcogen
D^0	Neutral chalcogen
d	Separation between metal islands
d_0	Initial island separation
$\bar{d}$	Average film thickness
E	Electric field
e	Magnitude of charge on an electron
$\mathscr{E}$	Energy
$\mathscr{E}_F$	Fermi energy
$\Delta\mathscr{E}$	Differential energy
F	Flux of oxidant
F'	Field strength of an ion
f	Fraction of preferred growth sites at dislocation ledges on a crystal surface
ΔG	Free energy change for crystal growth
ΔG_s^0	Single ion solvation free energy change
ΔH	Heat of adsorption
ΔH_d^0	Formation enthalpy for a defect pair in a network
h	Gas-phase transport coefficient for oxygen
h_f	Final oxide thickness
h_i	Initial oxide thickness
$\hbar$	Planck's constant over 2π
J	Current density
J_0	Constant
J_e, J_e'	Uncoupled tunneling current density
J_i	Ionic current density
K	Constant
K_H	Henry's law constant
K_n, K', K''	Constants ($n = 1, 2, 3, \ldots$)
k	Boltzmann constant
$k'f(x)$	Oxidation rate constant
k_c	Linear rate constant
k_p	Parabolic rate constant
k_0	Rate constant for zero-order reaction

k', k_1, k_1'	Logarithmic-type rate constants
k''	Inverse logarithmic rate constant
k_2, k_2'	Constants
L	Magnitude of oxide thickness at a specific time
L_0	Oxide thickness marking the transition from one space charge region in an oxide to another
L'	Latent heat for the glass transition
l	Length
$\mathscr{L}$	Grain size
M	Mass
m	Mass of a molecule
m'	Effective mass of an electron
N	Surface concentration
N_0	Number of Fermi electrons moving toward a metal–oxide interface
N_1	Number of oxidant molecules incorporated into a unit volume of oxide
n	Number
n'	Refractive index
n_c	Volume concentration
n_c^0	Concentration of bonds available for breaking
n_d	Volume concentration of defects
n_s	Number of surface sites per cm^2
n_v	Number of trapped electrons per cm^3 of oxide
P	Pressure
p	Corrosion penetration
Q	Charge
Q'	Activation energy
q	Magnitude of charge on an ion
R	Resistance
R_0	Initial film resistance
r	Radius of metal island
$r_{1,2}$	Charge separation distance for D^-
S	Solubility
ΔSP	Change in surface potential
T	Temperature

T_f	Crystallization temperature
T_g	Glass transition temperature
T_m	Melting point
T_3^+	Three-coordinated silicon
t	Time
t_i	Initial time
t_0	Constant
U	Activation energy which must be overcome by an ion moving from one equilibrium site in an oxide to an adjacent site
u	Oxide growth rate
u_1	Crystal growth rate per unit area
V	Voltage
V_k	Kinetic potential across a growing oxide film
v	Drift velocity of ions
W	Activation energy needed for moving an ion from the surface of an oxide into the interior (sum of $W_i + U$)
W_A	Energy barrier for anion movement
W_B	Binding energy of an oxygen ion to an oxide surface
W_C	Energy barrier for cation movement
W_i	Heat of solution of an ion in an oxide
W_m	Mass loss
W_t	Energy of an electronic trap in an oxide
W'	Activation energy for oxidant transport in fused silica
W''	Activation energy for electron transfer
w	Width
w'	Oxygen uptake
x	Variable used to denote oxide thickness
x_0	Thickness of space charge region
x_1	Oxide thickness where the energy for thermally driven diffusion equals that for field-driven drift
x_i	Initial oxide thickness
x_L	Limiting oxide thickness postulated for logarithmic-type oxidation
Y	Electron affinity
y	Interval between ions
Z	Number of charges on an ion

z	Oxygen uptake
α	Sticking coefficient
α'	Constant
β'	Constant
γ	Energy with which an electron is bound to the interstitial ion in an oxide
γ'	Constant
δ^-	Partial negative charge
δ^+	Partial positive charge
ϵ	Permittivity
ϵ_0	Permittivity of free space
ζ, ζ'	Constants
η	Viscosity
θ	Fraction of surface covered
θ'	K_1/xkT
κ	Dielectric constant
λ	Wavelength of light
λ'	Constant
Λ	Area
μ	Mobility
ν	Atomic vibration frequency
ν'	Frequency factor for transport of ions to a growing crystal surface
ν_i	Ionic attempt frequency
ρ	Resistivity
ρ_d	Dislocation density
ρ'	Oxide density
σ	Conductivity
σ_s	Stress
σ'	Metal–oxide thickness conversion factor
σ''	Surface charge density
Σ	Mass
τ	Constant
ϕ	Electronic energy barrier; the energy required to remove an electron from the Fermi level of a metal into the conduction band of an oxide
ϕ'	Si–O–Si bridging bond angle in SiO_2

SYMBOLS

ϕ_i	Energy required to ionize an impurity within an oxide
ϕ_v	Work function of a metal against vacuum
δ_ϕ	Electron work function change
χ_i	Longitudinal compressive stress
ψ-	Fraction of oxide remaining free of islands
Ω-	Oxide volume formed per diffusing species

LOW-TEMPERATURE OXIDATION

1

INTRODUCTION

Corrosion, tarnishing, and oxidation at low temperatures all refer to the same phenomenon, the reaction of an element with its environment to form a compound, for example, an oxide, sulfide, hydroxide, or halide.

Vitreous oxides fulfill a unique function in this process. They serve as protective films on metals and semiconductors, in that they slow the rate of ion movement during oxidation or corrosion. This is possible because of their structural perfection. At the same time, they provide a more perfect interface at an oxide-semiconductor junction, leading to better electronic properties of devices built in the semiconductor.

The process of oxidation can be divided conveniently into two regimes, high temperature and low temperature, depending on the metal and time–temperature-pressure relationships during oxidation. The products of high-temperature oxidation are often polycrystalline, and as such, contain paths of easy-ion diffusion. However, the possibility of obtaining the perfection of vitreous oxide offers a means for correcting this problem.

Two areas of oxide technology can be identified wherein vitreous oxides readily form. The first, which includes low-temperature thermal oxidation and solid, liquid, or gas-phase anodization, depends for its effectiveness on the field-dependent mechanism of growth. The second, oxidation of semiconductors at high or low temperatures, is more dependent on inherent material properties. These two subjects will serve to illustrate the theme of this book, the role of vitreous oxides in oxidation.

1.1. IMPORTANCE OF THIN OXIDE FILMS

Metals generally are unstable in the climatic conditions of Earth. Thermodynamically, only a metal as noble as gold should survive as a native metal, resisting conversion into an oxide, halide, sulfide, or other compound. The important metals used in engineering and construction, iron, aluminum, and copper, for instance, all corrode to varying degrees [1].

Thus, for long-term usefulness, they require either protective coatings or periodic renewal by metallurgical means.

The cost of replacing corroded metal is staggering. It was estimated in 1975 that 70 billion dollars were lost due to the corrosive effects of urban environments, chemicals, seawater, and other agents of corrosion [2]. Ten billion dollars of this amount were avoidable. The total continues to increase with the accelerated burning of fossil fuels. Hence, preventing the destruction brought about by corrosion is a very important goal [3] which has occupied scientific minds since before Faraday's time [4,5].

The destruction of metal by corrosion is the costly aspect of oxidation at low temperatures. Perhaps it will be surprising to some when they are reminded that useful purposes for the reaction called corrosion are also known. In cathodic protection of metals [4], the wasting away of one metal prevents the destruction of a more noble metal. Likewise, in zinc–air, aluminum–air, and iron–air [6] batteries, the reaction of a metal with oxygen is encouraged by the presence of alkali so that a reasonable electric current can be obtained through corrosion at room temperature. Metal etching might be considered corrosion pushed to the limit.

A thin (usually ~30 Å) stable oxide formed by oxidation at moderate temperatures, say below 250°C, plays a useful role, albeit hidden, in promoting lubrication [7]. Bare metal against bare metal produces a cold weld and as a result, high friction in a bearing. Oxide formation combined with metal ductility and lubrication all play a role in keeping the metal surfaces apart.

In like manner, thin oxide layers often influence the adhesion of nonmetals to metals. In the medical field, prosthetic devices [8,9] are placed in a very corrosive environment, the human body. They depend for their usefulness on the protective nature of the surface layers formed on the implant. In less corrosive environments, the bonding of glass or plastics to a metal can depend on the existence of an intermediate oxide layer [10].

Thin oxide films have also been found to affect the diffusion rate of gases out of metals, for example, tritium out of niobium [11].

An important role is played by oxidation at low temperatures in electronic components such as contacts [12] and integrated circuits [13]. Increase in contact resistance due to oxidation is a problem in integrated circuits [14], but thin oxide films, usually 1000 Å or less, also play a key role in the proper functioning of solid-state devices. The discovery of Atalla et al. [15] in 1959 that thermal oxidation of silicon stabilizes, that is, passivates, its surface was a crucial step in semiconductor device technology. Prior to this discovery, semiconductor devices based on germanium were plagued by surface instability effects which resulted from chemical interactions between the semiconductor and its environment. In fact, the development of integrated circuits, and thus, today's computer industry, is largely due to the excellent properties of the passivating SiO_2 film, including its interface with the silicon substrate. The importance of the noncrystalline structure of the SiO_2 film in this application has been pointed out by Revesz [16]. Revesz and Kruger [17], following Young [18], Hoar [19], and Fehlner and Mott [20], have also suggested that noncrystalline oxide films can play a decisive role in the passivation of metals. Although the passivating SiO_2 films on silicon are usually prepared at elevated temperatures (700–1200°C) and can be significantly thicker than the few score Angstroms characteristic of metal oxide films obtained at lower temperatures, they exhibit great similarity to thin oxide films, especially in their structure. Thermally grown SiO_2 films on single crystal silicon are vitreous, that is, noncrystalline, but with a high degree of short-range order. In fact, SiO_2 can be considered as a vitreous solid *par excellence*. For this and other reasons to be discussed in Chapter 3, the oxidation of silicon and the properties of SiO_2 films will be described as a model system for vitreous oxides. Since less understanding exists of noncrystalline solids than of crystalline solids, an extensive treatment of vitreous oxides is included.

Thin oxide films play an important role in both passive and active solid-state devices. Examples of the former are tantalum oxide capacitors prepared by anodization. Examples of the latter include tunneling junctions [21] and MOS transistors, especially charge storage [22] and tunneling MOS types [23]. Specialized devices such as thin film tunnel diodes and triodes have been proposed as active devices for integrated electronic circuits [24]. The presence of a thin oxide film between the semiconductor and metal in a Schottky barrier junction, as used in photovoltaic devices [25,26], has been found to improve performance of solar cells in converting light energy to electricity. The reverse process, conversion of electricity to light, can be achieved in a metal–insulator–metal tunneling junction [27]. Another thin film device, the cryotron, has been proposed for use in cryogenic memories [28]. However, the difficulties in fabricating reproducible, stable, oxide layers have slowed the utilization of such thin film devices.

Renewed interest in tunneling devices has been centered around the Josephson effect [29]. Josephson predicted that when a tunneling junction is made from superconducting metals and operated at temperatures below the superconducting transition temperature, then such junctions would exhibit both a DC threshold switching and an AC magnetic-field-dependent oscillation. Recent efforts [30,31,32] have adapted the Josephson junction to a computer memory cell which has a faster cycle time (~1 nsec) than silicon technology can achieve. One of the crucial steps in fabricating Josephson tunneling devices is the preparation of the thin oxide films on the superconducting metal film.

1.2. NATURE OF OXIDATION

The oxide films used in electronic devices as well as the passivating oxide films which prevent further corrosion of a metal or semiconductor are usually formed by oxidation. Oxidation, in a general sense, is the reaction of a metal (or semiconductor) with a nonmetal (chalcogenide, halogen, water, or combination) to produce a compound on the surface of the metal. Water can play a vital role in the reaction of a metal with oxygen, although dry oxidation will be stressed in the present work. Several recent studies discuss the corrosion aspects of low-temperature oxidation, especially as it applies to electronic circuits and components. They will be discussed further in Chapter 8.

A very intriguing change occurs during oxidation. The reactants, a metal or semiconductor having delocalized bonding and a gas having covalent bonding, are converted into a product, an oxide having partially ionic, partially covalent bonding. When the product of reaction is a solid, it separates the two reactants. Continuation of the reaction requires that a species of metal and/or oxidant dissolve in and move through the growing oxide. This species is often assumed to be ionic but there is evidence that, at least in the case of silicon oxidation, the migrating species can be molecular oxygen. To maintain charge neutrality during passage of a single ionic species, an electronic species (electron or hole) must also move through the oxide in the same or opposite direction.

Barriers to ion movement into and through the oxide exist. They are illustrated by the potential energy diagram in Figure 1.1 where U represents the energy which must be overcome by an ion moving from one equilibrium site in the oxide to an adjacent one separated by distance $2a$, while W is the sum of U plus W_i. Thus, W represents the energy change which occurs when an ion is removed from the oxide surface into the bulk of the oxide.

The means for overcoming these energy barriers form the basis for distinguishing low-temperature from high-temperature oxidation. In high-temperature oxidation, thermal activation is sufficient to account for ion generation and movement through the oxide even though a small electric field may be present. A parabolic rate law is generally followed. In low-temperature oxidation where the thermal energy is insufficient to allow existing ions or electrons to surmount the energy barrier, an alternative driving force such as an electric field or stress is postulated to account for continued reaction. Logarithmic-type kinetics are followed in this

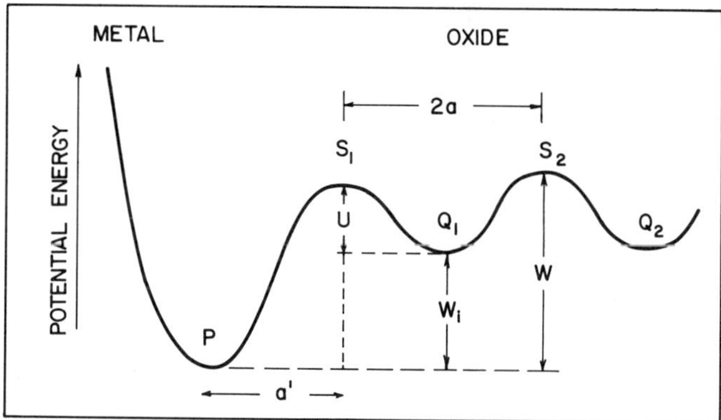

Figure 1.1. Potential energy diagram at a metal–oxide interface. *P* is the position of an ion at rest in the interface, *Q* is a lattice or network site in the oxide, and *S* is the top of the potential energy barrier separating the *Q* sites. W_i is the heat of solution of a metallic ion, *U* is the activation energy for diffusion in the oxide, and $W = W_i + U$. The jump distance a' required for an ion to enter the oxide is taken to be equal to half the ion jump distance within the oxide, that is, $a' = a$. After Mott [33]. Published by permission of the copyright holder, The Institute of Physics.

case. The actual temperature of transition from low- to high-temperature oxidation is a function of the metal involved, its perfection, and purity.

A brief description of the reactants and products involved in oxidation is appropriate at this point.

Metals make up a large portion of the periodic chart and vary greatly one from another in both physical and chemical properties [34]. Stone [35] has classified the oxidative properties of metals according to their atomic numbers. The distributed bonding in metals is perhaps of greatest pertinence to low-temperature oxidation. It stems from the free electron sea which surrounds the ionic nuclei and accounts for most metallic properties. It is this same electron sea that produces the nondirectional nature of the bonding in metals which must be overcome when a metallic atom dissolves in the adjacent oxide as an ion.

The crystallographic state of the metal substrate is of special interest. Single crystal, polycrystalline, and amorphous metals are found in practice. Single crystal and amorphous differ from polycrystalline in that no grain boundaries are present. Impurities often concentrate at grain boundaries, leading to defect regions in the oxide grown from the polycrystalline metal. These regions provide paths for easy-ion movement and thus, fast oxide growth. Single crystal and amorphous metals minimize such defects and should, therefore, produce more perfect oxides which result in a slower rate of oxidation [36].

Little need be said here about the reactant, oxygen. It is a diatomic gas which is thermodynamically favored to combine with metals to form oxides. Only kinetic parameters, determined in large part by the product oxide, prevent instantaneous, complete reaction. It is important to note that the reactivity of oxygen and/or mass transport through the oxide are often increased by the presence of impurities, especially water, in the oxidizing environment.

The oxide product of the reaction normally forms between the two reactants unless the oxide is volatile. Since the reactants must pass through the oxide, its properties with respect to electronic and mass transport are most important to continued reaction. These properties are intimately tied to the structure of the oxide film.

Oxides, like metals, occur in several forms: single crystal, polycrystalline, and vitreous. Again, the polycrystalline material differs in behavior from the other two because it contains grain boundaries which provide paths for easy-ion movement [37]. Ions move with greater difficulty through the lattice of single crystals or the network of vitreous oxides. Various aspects of oxide structure as well as electronic and mass transport in oxide films are discussed in Chapters 3 and 4.

It must be concluded that oxidation is a complex process at any temperature. Crystallography, mechanical defects, electronic properties, mechanisms of ion movement in the oxides, impurities, temperature, and oxygen pressure all play a vital role. At high temperatures, the interrelationships between these variables have been fairly well defined [38,39,40]. The present work begins to outline the relationships [20,41] which occur during the formation of vitreous oxides, especially during silicon oxidation and oxide formation at low temperatures.

REFERENCES

1. M. Pourbaix, *Lectures on Electrochemical Corrosion*, Plenum Press, New York, 1973.
2. N.B.S. Spec. Publ. 511-1, "Economic Effects of Metallic Corrosion in the United States," L. H. Bennett et al., 1978.
3. F. H. Haynie, in *Atmospheric Corrosion*, W. H. Ailor, Ed., John Wiley & Sons, New York, 1982, p. 3.
4. U. R. Evans, *An Introduction to Metallic Corrosion*, 2nd ed., Edward Arnold Ltd., London, 1963.
5. H. H. Uhlig, *Corrosion and Corrosion Control*, 2nd ed., John Wiley & Sons, New York, 1971.
6. L. Carlsson and L. Öjefors, *J. Electrochem. Soc.* **127** (1980), 525.
7. D. V. Keller, Jr., in *Surfaces and Interfaces I*, J. J. Burke et al., Eds., Syracuse University Press, Syracuse, NY, 1967, p. 225.

8. N. Ramasamy, B. R. Weiss, B. Stanczewski, P. N. Sawyer, and G. W. Kammlott, *J. Electrochem. Soc.* **123** (1976), 1662.
9. P. N. Sawyer, B. Stanczewski, N. Ramasamy, G. W. Kammlott, J. G. Stempak, and S. Srinivasan, *Trans. Amer. Soc. Artif. Intern. Organs* **19** (1973), 195.
10. F. M. Fowkes, in *Surfaces and Interfaces I*, J. J. Burke et al., Ed., Syracuse University Press, Syracuse, NY, 1967, p. 197.
11. D. Chandra, T. S. Elleman, and K. Verghese, *J. Nucl. Mater.* **59** (1976), 263.
12. R. S. Timsit, *Appl. Phys. Lett.* **35** (1979), 400.
13. S. M. Goodnick, M. Fathipour, D. L. Ellsworth, and C. W. Wilmsen, *J. Vac. Sci. Technol.* **18** (1981), 949.
14. E. H. Nicollian and J. R. Brews, *MOS (Metal-Oxide Semiconductor) Physics and Technology*, John Wiley & Sons, New York, 1982.
15. M. M. Atalla, E. Tannenbaum, and E. J. Scheibner, *Bell Sys. Tech. J.* **38** (1959), 749.
16. A. G. Revesz, *Phys. Stat. Sol.* **19** (1967), 193.
17. A. G. Revesz and J. Kruger, in *Passivity of Metals*, R. P. Frankenthal and J. Kruger, Eds., The Electrochemical Society, Princeton, NJ, 1978, p. 137.
18. L. Young, *Anodic Oxide Films*, Academic Press, New York, 1961, p. 171.
19. T. P. Hoar, *J. Electrochem. Soc.* **117** (1970), 17C.
20. F. P. Fehlner and N. F. Mott, *Oxid. Metals* **2** (1970), 59.
21. G. Unterkofler, *Electro-Tech.* **75** (1965), 55.
22. E. C. Ross, A. M. Goodman, and M. T. Duffy, *RCA Rev.* **31** (1970), 467.
23. W. E. Dahlke and S. M. Sze, *Solid-State Electronics* **10** (1967), 865.
24. C. A. Mead, *J. Appl. Phys.* **32** (1961), 646.
25. D. R. Lillington and W. G. Townsend, *Appl. Phys. Lett.* **28** (1976), 97.
26. J. Shewchun, D. Burk, R. Singh, M. Spitzer, and J. Dubow, *J. Appl. Phys.* **50** (1979), 6524.
27. R. K. Jain, S. Wagner, and D. H. Olson, *Appl. Phys. Lett.* **32** (1978), 62.
28. J. Matisoo, *Proc. IEEE* **55** (1967), 172.
29. B. D. Josephson, *Phys. Lett.* **1** (1962), 251.
30. W. Anacker, *IEEE Trans.* **MAG 5** (1969), 968.
31. H. H. Zappe and K. R. Grebe, *J. Appl. Phys.* **44** (1973), 865.
32. Various papers in the *IBM J. Res. and Dev.* **24** (1980), No. 2.
33. N. F. Mott, *Trans. Faraday Soc.* **43** (1947), 429.
34. E. A. Moelwyn-Hughes, *Physical Chemistry*, 2nd ed., Pergamon Press, New York, 1961.
35. H. E. N. Stone, *J. Mater. Sci.* **7** (1972), 1147.
36. J. C. C. Fan and V. E. Henrich, *Appl. Phys. Lett.* **25** (1974), 401.
37. D. R. Campbell, E. I. Alessandrini, K. N. Tu, and J. E. Lewis, *J. Electrochem. Soc.* **121** (1974), 1081.
38. O. Kubaschewski and B. E. Hopkins, *Oxidation of Metals and Alloys*, 2nd ed., Butterworths, London, 1962.
39. K. Hauffe, *Oxidation of Metals*, Plenum Press, New York, 1965.
40. P. Kofstad, *High Temperature Oxidation of Metals*, John Wiley & Sons, New York, 1966.
41. N. Cabrera and N. F. Mott, *Rep. Prog. Phys.* **12** (1948–49), 163.

2

RATE EXPRESSIONS FOR OXIDE GROWTH

2.1. INTRODUCTION

Thermodynamics states the conditions of temperature, pressure, and chemical composition under which an oxide can form on a metal. Kinetics complements thermodynamics in describing the rate of reaction, since in most cases transport of a reactive species through an oxide is necessary for oxidation to proceed. Thus, oxide structure is very important.

Thermochemistry is not covered in the present work since low-temperature oxidation relies on an electrochemical mechanism. However, there are abundant sources of thermodynamic data and discussions relevant to thermally driven oxidation. Books on high-temperature oxidation [1,2] and on glass [3] include tables of thermal data.

Evans [4] pointed out in 1960 that a host of kinetic expressions have been derived and/or measured to describe oxidation. Parabolic, linear, sigmoidal, and logarithmic laws are encountered. In addition, combinations resulting from the transition of one growth stage to another are possible. These tend to be higher order expressions, for example, cubic [5]. Thus far, no single kinetic expression has been found which describes the complete course of oxide growth on a clean metal surface at all temperatures and pressures.

The variety in rate expressions is convenient, but confusing at times. Similar rate expressions can be found for different temperature ranges or oxide thicknesses and are explained by totally different mechanisms. For instance, a logarithmic dependence of oxide film thickness x on time t can be found for compact films a few tens of Angstroms thick, and also for thicker films at higher temperatures where cavities form at the interface between metal and oxide. Likewise, the parabolic expression often describes oxidation at high temperatures, but can be encountered at lower

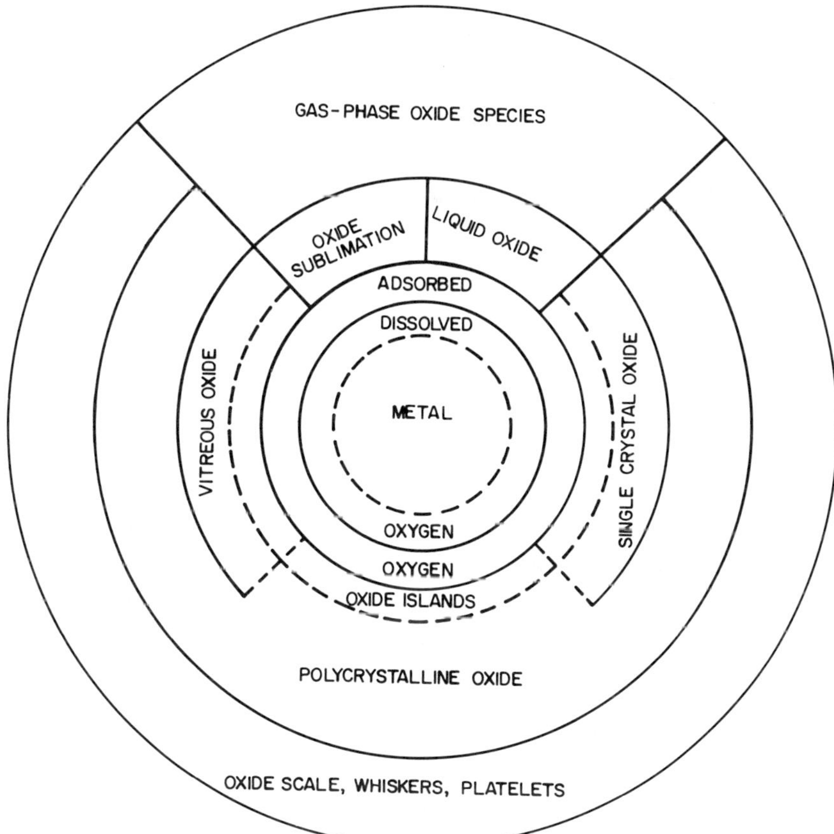

Figure 2.1. Diagram showing the evolution of oxide phases formed during oxidation of a metal. Time and temperature increase radially outward from the center.

temperatures where the velocity of ions moving through the oxide varies directly with a small electric field.

Increasing the confusion is the fact that it is sometimes difficult to distinguish one rate law from another by simply plotting the experimental data. This is especially true for the case of logarithmic expressions which include direct logarithmic, inverse logarithmic, and asymptotic expressions.

Possible oxide structures are shown schematically in Figure 2.1, where time and temperature, that is, extent of reaction, increase radially outward. Each succeeding shell implies a new rate-limiting process such as interface reaction, limitations due to the bulk oxide, or the requirement for an adequate oxygen supply. Impurities in particular are critical in oxidation.

Assumptions are often made in constructing models of oxidation processes. Charge neutrality requires that ionic movement be balanced by movement of a species of opposite charge, usually electronic, through the oxide. It is often convenient to assume that one of the processes dominates. The faster process is assumed to be in a virtual equilibrium while the slower one is assumed to be rate determining. This is the so-called single-current approach to oxidation models.

A more exact approach is to consider all processes simultaneously so that the interactions between them can be determined. This approach forms the heart of the coupled-currents approach as practiced by Fromhold [6], Dignam [7], and others. It is made possible by numerical integration using high-speed computers.

2.2. SINGLE-CURRENT CONTROL OF OXIDATION—OXIDATION LIMITED BY BULK OXIDE

Limitation of oxide growth rate by processes occurring in the oxide itself has been the most common assumption in evaluating growth kinetics. Various entities can move through an oxide, for example, atoms or molecules, both charged and uncharged, electronic species, and impurities. Oxide structure, especially the presence of pores and grain boundaries, can greatly affect the relative mobilities of these species.

The movement of a neutral species is described by a simple transport expression based on a steady-state approximation. A chemical gradient drives the reaction. Thus,

$$J = \frac{D[C(O) - C(L)]}{x(t)} \tag{1}$$

where J is the current of neutral species,
D is the diffusion coefficient,
C is the concentration at interfaces O and L, and
L is the value of oxide thickness x at time t

This process may occur in pores or grain boundaries in the oxide. Given $dx(t)/dt = \Omega J$ where Ω is oxide volume formed per diffusing neutral species, this single-current approach to metal oxidation yields upon integration a parabolic growth law of the form

$$[x(t_L)]^2 - [x(t_o)]^2 = 2k_2't \quad (2)$$

where k_2' is $\Omega D[C(O) - C(L)]$. Equation 2 represents a basic expression derived by early workers in the field of oxidation.

An exponential temperature dependence for k_2' is expected since D and often C follow the Arrhenius expression. The pressure dependence of k_2' is related to the concentration of diffusing species. It varies in crystalline oxides from almost no dependence in the case of metal interstitials or oxygen vacancies to varying as $P^{1/2n}(1 \leqq n \leqq 4)$ for oxygen interstitials or metal vacancies. In vitreous SiO_2, k_2' varies as P and is related to the molecular solubility of oxygen in SiO_2.

The diffusion of ionic species through an oxide is subject to the influences of an electric field. Hence, the current J of mobile reactants becomes, in the presence of a field

$$J = -D\frac{\partial C(x,t)}{\partial x} + \mu EC(x,t) \quad (3)$$

Here E is the electric field and μ the mobility of the charged species, related to the diffusion coefficient D by the Einstein relationship

$$ZeD = \mu kT \quad (4)$$

Equation 3 is the well-known Wagner relationship. Fromhold [6] and Hauffe [8] treat several specific cases of this equation.

For $x > 100$ Å and $E \leqq 10^5$ V cm^{-1}, Fromhold has integrated Equation 3 for the steady-state condition

$$\frac{\partial J}{\partial x} = 0 \quad (5)$$

to give

$$J_s = \mu_s E\left[\frac{C_s(L) - C_s(O)\exp(\mu_s EL/D_s)}{1 - \exp(\mu_s EL/D_s)}\right] \quad (6)$$

for species S. Fromhold treats specific cases in which the voltage V across the film and the particle currents are derived based on the coupled-currents criterion

$$\sum_{s=1}^{r} Z_s e J_s = 0 \tag{7}$$

Evans [4] has also derived a general expression for ion movement through an oxide under the influence of a gradient (electric or chemical) which varies as $1/x$. He finds that, based on the change in probability of movement due to the imposed gradient,

$$\frac{dx}{dt} = A'K[e^{\theta'} - e^{-\theta'}] \tag{8}$$

where $A' = \exp[-W/kT]$
$\theta' = K_1/xkT$

K and K_1 are constants while W is the barrier to ion transport as shown in Figure 1.1.

The expansion of Equation 8 is

$$\frac{dx}{dt} = 2A'K\left[\theta' + \frac{\theta'^3}{3!} + \frac{\theta'^5}{5!} + \cdots\right] \tag{9}$$

so that at large x the higher order terms in θ' can be neglected leading to the parabolic expression

$$\frac{dx}{dt} = 2A'K\theta' = \frac{K_2}{x} \tag{10}$$

where

$$K_2 = \frac{2A'KK_1}{kT}$$

Upon integration

$$x^2 = 2K_2t + K_3 \tag{11}$$

An expression for small x is presented below.

Anderson and Ritchie [9] have also derived a general oxidation expression by applying random walk theory to tarnishing reactions. A pseudo three-dimensional model delineates conditions under which linear, parabolic, or logarithmic kinetics predominate. The effect of an electric field is included.

Cabrera and Mott [10], in a general treatment of oxidation, derived a parabolic rate law for oxidation limited by cation diffusion in the bulk oxide,

$$x^2 = 4\Omega D n_c t \tag{12}$$

where Ω is the oxide volume per metal ion, n_c is the concentration of ions at the metal–oxide interface, and D is the diffusion coefficient for ions. The concentration of metal ions at the gas–oxide interface was assumed to be zero.

Cabrera and Mott adopted a criterion to distinguish this parabolic growth based on bulk oxide control from other modes of growth based on interface control. It utilizes the thickness x_0 of the space charge region induced in the oxide by the metal or semiconductor contact.

$$x_0 = \sqrt{\frac{\kappa k T}{8\pi n_c e^2}} \tag{13}$$

where κ is the dielectric constant of the oxide and n_c is the concentration of dissolved ions. When $x > x_0$, the parabolic regime of growth is encountered. This is called high-temperature oxidation due to the need for thermally activated ions.

The value of n_c in Equations 12 and 13 is dominated by an activation energy, $W_i + \phi$, where W_i is the ionic heat of solution and ϕ is the energy required to remove an electron from the Fermi level of the metal into the conduction band of the oxide. Since ϕ is often of the order of 1 eV, $W_i + \phi$ is greater than 1 eV. As a result, temperatures of several hundred degrees Celsius are needed to produce a sufficient ionic population to support oxide growth. Parabolic growth is essentially driven by the concentration gradient of ions in thick oxides.

For x less than x_0 (Equation 13), the thin film case of oxide growth is considered in which the ions and electrons moving across the oxide layer can be considered independently, the space charge due to the dissolved ions can be disregarded, and a quasi-equilibrium of electrons is established between the metal and adsorbed oxygen. Films considered in this category are in the range of several hundred Angstroms or less. Two cases are distinguished by a second criterion which is energetic in nature. It is basically the ratio of energy from an electric field to that from thermal motion. This ratio is taken to be one, so that a thickness x_1 is defined

$$x_1 = \frac{qaV}{kT} \tag{14}$$

When $x < x_0$, but $x > x_1$, the first thin film case is found in which the diffusion velocity of ions in the oxide film is directly proportional to the field V/x, the condition $qaE < kT$ being satisfied. Under these conditions

$$\frac{dx}{dt} = \frac{n_c \mu V \Omega}{x} \tag{15}$$

Integration yields a parabolic rate law

$$x^2 = 2\Omega \mu n_c V t + c \tag{16}$$

Making use of the Einstein relationship $\mu kT = ZeD$, we find that Equation 16 becomes

$$x^2 = 2Zn_c D\Omega \left(\frac{eV}{kT}\right) t + c \tag{17}$$

Note specifically that Equation 17 (thin-film case) is quite different from Equation 12 (thick-film case). Equation 15 can also lead to linear or cubic rate laws when it is integrated, depending upon boundary conditions. The cubic rate law has also been explained in terms of a transition state by Evans [4].

The second thin-film case treated by Cabrera–Mott ($x < x_0$, $x < x_1$) is invoked when the field is so strong that the ion drift velocity is exponentially proportional to the field. This occurs because the field lowers the activation energy for ion movement in one direction, but increases it in the opposite direction, given $qaE > kT$. Fromhold and Cook [11] have estimated that kT/qa is $\sim 10^6$ V cm^{-1} at 300°K. Thus, E must be greater than this value.

For thin oxides under high fields (low-temperature oxidation), there is no interface equilibrium of ions and oxidation is controlled by either defect generation at the interface or by bond breaking within the oxide [12]. A quasi-equilibrium of electrons is also assumed in this case. In addition to

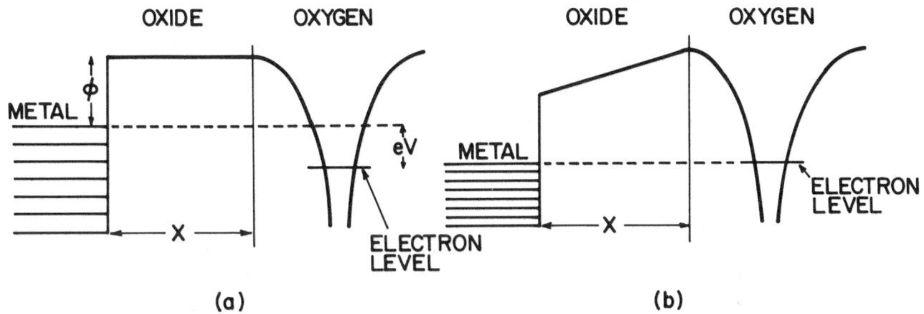

Figure 2.2. Electronic energy levels in the metal, oxide, and adsorbed oxygen: (a) before electrons have passed through the oxide; (b) when equilibrium is achieved. After Cabrera and Mott [10]. Published by permission of the copyright holder, The Institute of Physics.

Cabrera and Mott [10], Mosley and Fromhold [13] and Young and Dignam [14] have published analyses of this problem.

The essence of the Cabrera–Mott treatment for low temperatures is the hypothesis that adsorbed oxygen produces electronic trap sites at the oxide–oxygen interface and that electrons can freely pass through the oxide to populate these sites. This may occur, for example, by tunneling, impurity conduction, or Schottky emission into the conduction band of the oxide. As shown in Figure 2.2, the contact potential difference V thus set up between the metal and adsorbed oxygen creates a field E such that $E = V/x$.

The thin oxide on the metal is assumed by Cabrera–Mott to be crystalline, although highly compressed. The rate-determining step is then taken to be incorporation of a cation, typically from a kink site of the metal (see Figure 2.3) into an interstitial site of the oxide, requiring an energy W where $W = W_i + U$ as shown in Figure 1.1.

However, it is now known [15] that many oxides grown at low temperatures are noncrystalline and the concept of an interstitial cation may no longer be applicable. In addition, anions as well as cations can be mobile in anodic oxides. The possible lack of a rigid reference framework for ion movement in vitreous oxides complicates the description of this motion, as well as contributing to a change in structure with time. Ion movement in vitreous oxides is discussed in Chapter 4.

The approach used to derive the parabolic rate laws is no longer valid for such strong fields since equilibrium at the interfaces is not achieved. As a result, an approach similar to Equation 8 of Evans was adopted. Consider for a moment Figure 1.1. In the absence of a field, an ion has

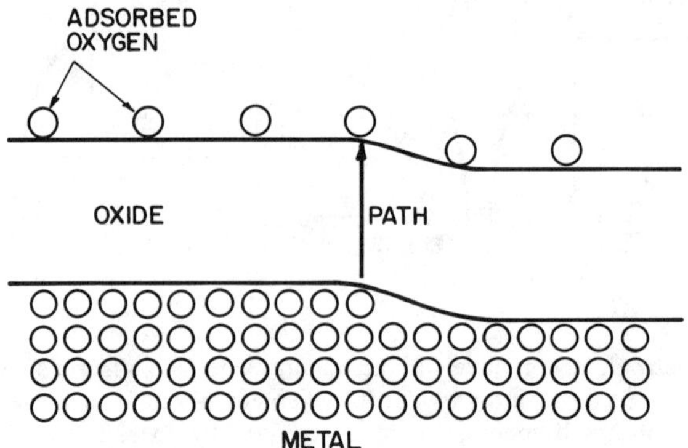

Figure 2.3. Proposed mechanism by which cations leave a step on the metal surface and pass through the oxide layer. After Cabrera and Mott [10]. Published by permission of the copyright holder, The Institute of Physics.

a probability of surmounting the potential barrier U and moving a distance $2a$ from interstitial site to interstitial site

$$\nu \exp\left[\frac{-U}{kT}\right] \tag{18}$$

where ν, the atomic vibration frequency, is taken to be 10^{12} $\sec^{-1}$. The presence of a field lowers the barrier U by qaE in the direction of the field and raises it by the same amount in the opposite direction. The drift velocity v of ions thus becomes $2a$ times the sum of the probabilities of motion in the two directions. For small E, the resulting expression can be reduced to v directly proportional to E, similar to Equation 15. For large values of E, the expression becomes

$$v \cong 2\nu a \exp\left(-\frac{U - qaE}{kT}\right) \tag{19}$$

It is this case which specifically pertains to oxidation at low temperatures and the range of very thin oxide films less than 50 Å thick.

Local ionic equilibrium can be ignored when the field is very strong. There is little or no time for ion exchange between the metal and the oxide so that a mobile ion, once inside the oxide, quickly passes to the other side. If this were not so, a higher voltage V due to stronger binding within the oxide would be encountered.

It appears then that conditions at the interfaces rather than in the oxide itself determine which ion will be mobile and under what conditions. To illustrate this, let us again examine the potential energy diagram of a metal ion at the metal–oxide interface, as in Figure 1.1. The total barrier to ion movement at the interface is W. Assuming that the first jump distance of the ion is equal to half the distance between interstitial sites in the oxide, we find that the probability that an ion will cross the interfacial barrier, and hence the oxide film is

$$\nu \exp\left[-\frac{W}{kT}\right] \exp\left[\frac{qaE}{kT}\right] \tag{20}$$

The rate of oxide growth thus becomes

$$\frac{dx}{dt} = N\Omega\nu \exp\left[\frac{-(W - qaE)}{kT}\right] \tag{21}$$

where W is the barrier to ion incorporation at the interface. The mathematical form of Equation 21 is similar to that of Equation 8.

To reiterate, no local equilibrium is considered to exist between oxide and metal so that an ion, once incorporated in the oxide, is pulled immediately across to the other side. The symbol Ω is the oxide volume per ion and N is the number of potentially mobile ions at the oxide interface. The quantity N may be related to the surface states so often discussed in the theory and practice of semiconductor technology [16] although no direct experimental connection between them has yet been established.

It is possible to extend the above treatment to the oxide–gas interface, provided anions are the mobile species as pointed out by Fehlner and Mott [15].

Equations 15 and 21 are actually special cases of the general equation,

$$\frac{dx}{dt} = a\nu \exp\left[-\frac{W}{kT}\right] \sinh\left(\frac{qaE}{kT}\right) \tag{22}$$

Young's book [17] on anodization shows that Equation 21 is equivalent to the basic expression put forward to describe anodic oxidation. A comparison of these two expressions offers one way in which the term, low temperature oxidation, can be defined in terms of experimental parameters.

During anodic oxidation, the ionic current J depends on the field E according to

$$J = \gamma'' \exp[\beta E] \tag{23}$$

where γ'' and β are evaluated by comparison of Equation 23 with Equation 21.

$$\beta = \frac{qa}{kT} \tag{24}$$

$$W = kT \ln \left(\frac{N\Omega\nu}{\gamma''} \right) \tag{25}$$

Experimentally, γ'' and β can be determined by applying Equation 23 to data from the anodization of a metal or semiconductor. Thus, values of W and a can be found.

Given the logarithmic nature of the oxidation process, Cabrera and Mott define a limiting thickness x_L based on the growth rate slowing to one atom layer in 10^5 sec. Thus,

$$x_L \cong \frac{Vaq}{W - 39kT} \tag{26}$$

A critical temperature $W/39k$ is thereby defined. At temperatures below the critical temperature, the film grows rapidly to the limiting thickness x_L and then essentially stops growing. At higher temperatures, there is no limiting thickness and a transition to linear (see below) or parabolic growth occurs. A value of critical temperature based on this assumption has been estimated to be 300°C for aluminum [10], 500°C for silicon [18], and 200 to 300°C for iron depending on surface conditions [4].

Cabrera and Mott give an approximate integration of Equation 21 as does Hauffe [8]. The result is

$$\frac{x_1}{x} = K'' - \ln t \tag{27}$$

where $x_1 = qaV/kT$ and K'' is a constant. This is the well-known Cabrera–Mott inverse logarithmic law for oxide growth at low temperatures. Unfortunately, it has been practically impossible to plot experimental data according to Equation 27 and distinguish it graphically from the direct logarithmic law of Evans.

Ghez [19] carried out a more exact integration of Equation 21 so that the validity of the theory could be better tested by fitting experimental data to predicted parametric relationships. He found an expression which is no longer dependent on an arbitrary x_L. Instead

$$\frac{x_1}{x} = -\ln\left(\frac{t+\tau}{x^2}\right) - \ln(x_1 u) \tag{28}$$

where τ is a constant,

$$u = N\Omega\nu \exp\left(-\frac{W}{kT}\right) \tag{29}$$

and

$$x_1 = \left|\frac{ZeaV}{kT}\right| \tag{30}$$

Equation 28 is a better representation of the Cabrera–Mott theory than Equation 27. The constant τ may be neglected if it is small compared to t. A plot of $1/x$ versus $\ln[t/x^2]$ should yield a straight line. Actual physical constants can then be calculated from the slope $-x_1^{-1}$ and intercept $-x_1^{-1}\ln(x_1 u)$. If necessary, τ can be determined by computer linearizing plots of $1/x$ versus $\ln[(t+\tau)/x^2]$.

The most exact formulation of the Cabrera–Mott model is found in the numerical integration of coupled ionic and electronic currents, to be discussed below.

Evans [4] utilized Equation 8 in deriving an expression for oxidation at low temperatures. He assumed that the term in Equation 8 representing ion movement against the field could be neglected, giving rise to

$$\frac{dx}{dt} = A'Ke^{\theta'} \tag{31}$$

This equation, given suitable assumptions, reduces to

$$\frac{1}{x} = \frac{1}{x_i} - K'\ln[c'(t-t_i)+1] \tag{32}$$

an inverse logarithmic expression where x_i is the initial oxide thickness at t_i.

2.3. ROLE OF OXIDE STRUCTURE

The oxidation models discussed above are applicable to compact, well-ordered oxides. However, in practice, oxides often crack and spall off the metal due to differences in crystal lattice dimensions and thermal expansion coefficients. For these reasons, pores, cavities, and other defects must be considered.

Reactant movement through pores can lead to a logarithmic expression. Such pores can exist as an intrinsic part of the oxide, resulting from grain boundary interactions or dislocation in the oxide. Evans [4] has considered two basic cases: where the pores are self-blocking and where adjacent pores interact (mutual blocking). The first case leads to an asymptotic expression

$$x = K_6[1 - \exp(-K_7 t)] \tag{33}$$

provided $x = 0$ at $t = 0$, while the second leads to a form of the direct logarithmic law

$$x = K_8 \log(c''t + 1) \tag{34}$$

Cavities at the oxide–metal interface can arise from coalescence of cation vacancies which are unable to diffuse from the oxide into the metal. These obstructions to ion movement can grow with time, affecting the overall rate of oxide growth. A rate law of the form

$$x = K_9 \ln(K_{10} t) \tag{35}$$

has been proposed by Evans to describe this case. In Equations 33–35, c'', K_6, K_7, K_8, K_9, and K_{10} are constants.

2.4. TWO-STAGE LOGARITHMIC LAW—SPACE CHARGE CONTROL

A buildup of two space charge regions in the oxide has been shown by Uhlig and coworkers [20,21,22] to lead to a two-stage logarithmic law of the form,

$$x = k_1 \log\left(\frac{t}{\zeta} + 1\right) \qquad \text{(first stage)} \tag{36}$$

$$x - L_0 = k_1' \log\left(\frac{t}{\zeta'} + 1\right) \qquad \text{(second stage)} \tag{37}$$

where L_0 is the oxide thickness marking the transition from one space charge region to the other. This model of oxidation has been criticized by Fromhold [23,24] and restated by Ritchie [25]. It appears to fit the oxidation kinetics of several metals [26,27,28,29]. Fehlner and Mott [15] have proposed that the dual logarithmic behavior results from reordering of the oxide accompanied by grain growth. Young and Dignam [28] take a different approach and interpret two-stage direct logarithmic growth of zinc in terms of the Cabrera–Mott theory as expressed by numerical integration. Resolution of the question requires a careful study of oxide morphology and its correlation with kinetics plus numerical integration techniques to better compare data with theory. Further discussion of the two-stage law can be found in Chapter 7.

The model of Uhlig has been simplified by Williams and Hayfield [30] who assumed that only one uniform space charge region existed in the oxide. Their equation may be expressed as

$$\frac{dx}{dt} = \alpha'' \exp(-\beta'' x) \tag{38}$$

$$\alpha'' = N_0 \Omega \exp\left[\frac{-e(\phi + 4\pi a_d n_s e)}{\kappa k T}\right] \tag{39}$$

$$\beta'' = \frac{4\pi a_d n_v e^2}{\kappa k T} \tag{40}$$

where N_0 = number of Fermi electrons moving toward the metal–oxide interface
ϕ = metal–oxide electron barrier
n_s = number of surface sites per cm^2
n_v = number of trapped electrons per cm^3 of oxide
a_d = average length of individual dipoles (the order of atomic spacing)
κ = dielectric constant
e = electronic charge

Equation 38 can be integrated to give

$$x = \frac{\ln \alpha''}{\beta''} + \frac{\ln \beta''}{\beta''} + \frac{\ln t}{\beta''} \tag{41}$$

Chattopadhyay [31] obtained a similar equation using rectifier theory of a metal–semiconductor junction.

2.5. ZERO-ORDER LAW—GAS KINETIC CONTROL

Oxide growth rate is independent of oxide thickness when the oxygen supply to the reacting surface is rate limiting, as found during the formation of an adsorbed monolayer. The basic equation controlling this process is the gas kinetic expression for the number of gas molecules hitting a surface per unit time

$$\frac{P}{\sqrt{2\pi mkT}} \tag{42}$$

where P is oxygen pressure and m is the weight of a molecule. Thus, the rate of oxide growth becomes

$$\frac{dx}{dt} = k_0 P \tag{43}$$

where k_0 is a constant inversely dependent on the square root of gas temperature.

2.6. SURFACE-CONTROLLED REACTIONS

Ritchie and Hunt [32] have analyzed the role of surface reactions and the influence of pressure on oxidation kinetics. They make the simplifying assumption that the rate-determining step is much slower than the others, which are taken to be at equilibrium rather than steady state. In addition, each oxygen species is assumed to occupy one surface site, that is, there is no double occupancy.

A linear rate law was found when the electron concentration was taken to be independent of oxide thickness. This was true for all possible surface reactions. Pressure dependence was predicted to be first power when molecular species are involved in the slow step and one-half power if the slow step follows dissociation. A linear rate law can also be derived if the electron concentration follows a Boltzmann distribution.

Parabolic kinetics in which the rate varies as log P_{O_2} are consistent with surface-rate-controlling steps, again where electron concentration is given by the Boltzmann distribution. A cubic growth law can also be derived. In this case, reaction rate is expected to decrease with increasing pressure.

A direct logarithmic law is found when the rate-controlling step is addition of an electron, yielding a charged species, that is,

$$O^- + e^- \rightarrow O^{2-}$$

This rate law is predicted to be independent of pressure [33].

The equations are somewhat complex and expressed in terms of Langmuir functions. The reader is referred to the literature for their exact form.

2.7. LINEAR LAW—INTERFACIAL REACTION CONTROL

Incorporation of oxide components at the reaction interface can become the rate-limiting process if ion incorporation and movement through the oxide is easy, that is, the activation energy W in Figure 1.1 is less than that for reaction at the opposite interface. In this case, the rate of oxide growth is independent of oxide thickness.

Control of oxidation by reaction at an interface usually leads to the linear or rectilinear law with some exceptions [25]. This is more generally expressed in the mixed parabolic equation [4]

$$\frac{dx}{dt} = \frac{k_p(C_o - C_i)}{x} \tag{44}$$

$$\frac{dx}{dt} = k_c C_i \tag{45}$$

for partial control by an interface reaction under steady-state conditions. Integration yields

$$\tfrac{1}{2}k_c x^2 + k_p x = K't + K'' \tag{46}$$

where

$$K' = k_p k_c C_o \tag{47}$$

For large t and x, the parabolic term dominates. For small t and x, the linear term dominates. Linear oxidation rates occur when the reaction rate is slower than the rate of transport via, for instance, cracks, channels, or pores in the oxide. These provide a path for oxygen flow to the oxide–metal interface.

Dependence of the parabolic rate constant k_p on temperature and pressure has already been discussed for Equation 2. Dependence of the linear rate constant k_c on pressure and temperature is complicated and varies with the reaction mechanism, for example, oxygen dissociation on suitable interface sites or cation release from step sites on the metal surface.

2.8. SIGMOIDAL LAW—CRYSTALLIZATION OF THE OXIDE

Island nucleation and lateral growth in a vitreous oxide film growing on a metal produce sigmoidal kinetics. Here, oxide thickness x as a function of t has an initial low rate which speeds up upon nucleation of the islands.

Completion of crystal lateral growth causes the oxidation rate to slow again.

Statistical methods have been applied to derive expressions for the fraction ψ of oxide remaining free of islands [4]

$$\psi = \exp[-Kt^n] \tag{48}$$

where n depends upon whether a constant density of nuclei is assumed ($n = 2$) or whether continuous nucleation is allowed ($n = 3$). The result in both cases is an S-shaped curve of ψ versus t, resulting in the rate equation

$$\frac{dx}{dt} = K'[1 - \exp(-Kt^n)] \tag{49}$$

2.9. ELECTRONIC RATE CONTROL

Electronic rate control of oxidation at low temperatures was proposed quite early by Mott [34]. The passage of electrons through the oxide film can be rate limiting provided the ion current is in a virtual equilibrium. For thin oxides where electron tunneling dominates, the number of electrons having energy $\mathscr{E}$ passing a barrier of height ϕ and width x is

$$A_3 \exp\left[\frac{-2x\sqrt{2m'(\phi - \mathscr{E})}}{\hbar}\right] \tag{50}$$

$\hbar$ being Planck's constant over 2π. Thus

$$\frac{dx}{dt} = K_4 \exp(-K_5 x) \tag{51}$$

and for initially clean metal surfaces,

$$x = K_{11} \log(ct + 1) \tag{52}$$

the direct logarithmic law. K_4, K_5, and K_{11} are constants.

2.10. COUPLED-CURRENTS APPROACH

Most of the rate laws discussed heretofore assume that the ionic current is rate limiting. Fromhold [6] maintains that the coupled-currents approach is a more exact solution to modeling oxidation phenomena since continued interaction of both electronic and ionic currents is allowed. The approach takes advantage of high-speed computers to gain accuracy. Instead of resorting to approximations for integrating the differential equations representing ionic and electronic currents, numerical integration of the expressions is carried out. However, assumptions are made that the experimental setup fits the calculated model. Hence, caution must be exercised in the application of the coupled-currents approach. It is only as good as the assumptions made about the processes occurring in the oxide. Oxide imperfections and contaminants can easily overwhelm the theoretically correct expressions for charge transport in a perfect oxide film.

The coupled-currents approach to oxidation theory has been applied by Fromhold [6] to two cases of particular interest in the present work. For the low-temperature case, he treats electron transport by tunneling or thermionic emission combined with ionic diffusion in homogeneous electric fields. Comparison is made with the original model of Cabrera and Mott, to which is added an ionic concentration gradient, an ionic equilibrium potential, and the possibility of a virtual current equilibrium (quasiequilibrium) for either charged species.

An analysis based on the tunneling current includes both the forward and reverse electronic currents as well as the ionic current. Some results of this coupled-currents analysis (film thickness and kinetic potential as a function of time with the potential across the oxide as a variable) are shown in Figure 2.4. A self-limiting expression is evident, similar to the logarithmic laws discussed above.

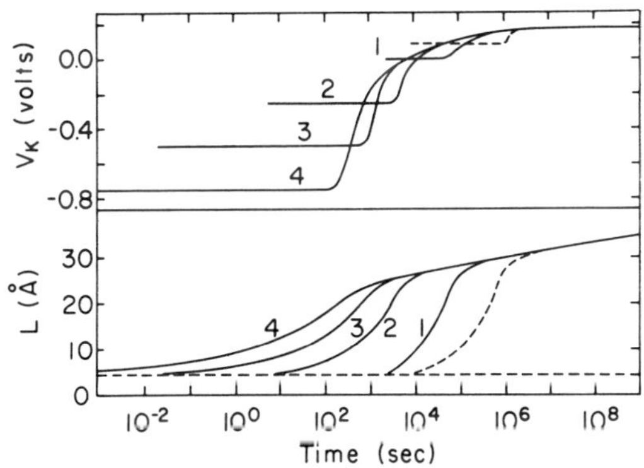

Figure 2.4. Film thickness L and kinetic potential V_k versus logarithm of time for different values of the self-induced potential V. Curves 1–4; $V = 0.00$, -0.25, -0.50, and -0.75 V respectively. For the dashed curve, $V = +0.10$ V. After Fromhold [6]. Published by permission of the copyright holder, North-Holland Physics Publishing.

The Hauffe–Ilschner [35] transition from electronic to ionic virtual current equilibrium is shown to occur at less than a monolayer in Figure 2.5. Experimental detection of the transition thus appears impossible. A second transition at essentially the thickness limit for tunneling represents a shift from ionic to electronic rate control. Comparison with the Cabrera–Mott theory indicates that a more complete rendition of events is given by the coupled-currents approach. However, determining whether oxidation experiments ever proceed beyond the ionic rate-limiting stage requires comparison of the equations with experimental data.

Electron tunneling is replaced by thermal electron emission (Schottky emission) at higher temperatures. Thickness–temperature regimes of applicability for these two electronic transport mechanisms, tunneling and thermionic emission, are shown in Figure 2.6.

Fromhold has calculated the example of thermionic emission combined with ionic diffusion in homogeneous electric fields. Results of numerical integration for this coupled-currents approach, assuming a virtual ionic equilibrium, give rise to an electric field which aids the Schottky emission. At higher temperatures, this ionic current equilibrium is destroyed and a virtual electronic equilibrium dominates, causing the potential across the

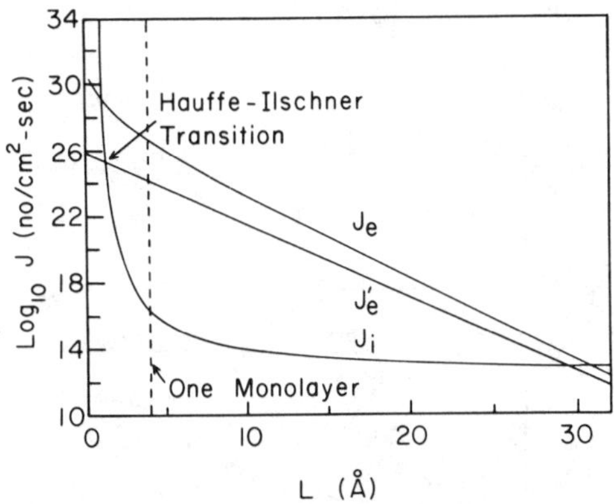

Figure 2.5. Uncoupled tunneling (J_e and J_e' calculated using different values of Richardson's constant) and ionic currents J_i versus film thickness L. Values for the parameters used to calculate the values of J were $V = -0.5$ V, $\phi = 1.0$ eV, $W = 0.80$ eV, $N = 4 \times 10^{10}$ ions cm^{-2}, $\nu_i = 10^{15}$ sec^{-1}, $T = 300$°K, and $2a = 4$ Å. Equation 21 was used to calculate J_i while an equation similar to Equation 50 was used for J_e. After Fromhold [6]. Published by permission of the copyright holder, North-Holland Physics Publishing.

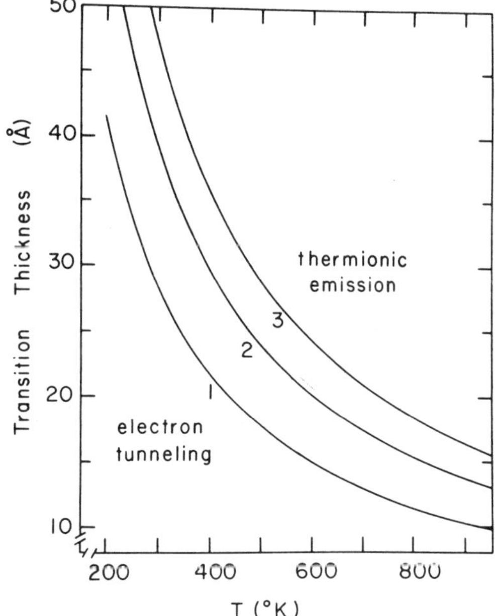

Figure 2.6. Temperature dependence of the transition thickness of dielectric at which electron thermionic current is equal to electron tunnel current. Metal–oxide electron work function $\phi = 0.5$, 1.0, and 1.5 eV for curves 1, 2, and 3, respectively. After Fromhold [6]. Published by permission of the copyright holder, North-Holland Physics Publishing.

oxide to reverse in sign. Growth kinetics tend to follow the expression

$$x(t) \propto t^{1/n} \qquad (1 \leqq n \leqq 4) \tag{53}$$

where the value $n = 2$, that is, the parabolic law, is often found.

Space-charge fields in the oxide can perturb the flow of charged species through the oxide. This is not considered to be important for thin oxides where the space-charge layer is greater than the oxide thickness. In such cases, a homogeneous electric field is assumed. However, Fromhold does calculate examples of the effect which space charge can have on the distribution of ionic and electronic species in a growing oxide. These calculations have been extended to include oxide thickness variations, electric field across the oxide, and the role of space charge in oxide kinetics. This latter effect is interesting, for space charge can enhance or retard the growth rate depending on the nature of the space charge and the sign of the rate-limiting species. Such effects appear significant for films thicker than 10 monolayers (~50 Å) having charge densities greater than 10^{17} cm^{-3}.

Dignam et al. [14,28,36] have constructed a set of equations which considers both concentration gradient and space charge due to moving ions. Their results agree in many respects with those derived by Fromhold and Cook [11], but emphasize the role of interfacial processes. Some highlights of the treatment of Dignam et al. are as follows. Space-charge effects are predicted to be significant at concentrations exceeding 5×10^{19} cm^{-3}. Boundary conditions are allowed to vary with film thickness during oxidation. The pressure dependencies of the predicted rate laws, for example, inverse logarithmic, cubic, parabolic, and linear, are included. Direct logarithmic kinetics are possible for both cation vacancy and cation interstitial models [37].

2.11. ROLE OF STRESS

Stress is an additional factor that may have an influence on diffusion in solids and thus oxidation. However, predicting and evaluating the effects of stress can be quite difficult. Isostatic pressure increases the energy needed to form vacancies within a material and thus makes diffusion more difficult [38]. For self-diffusion by a defect (vacancy) mechanism, an increase in pressure decreases the rate of diffusion.

The situation is entirely changed when a biaxial stress is applied to a material. In this situation, vacancies tend to flow from source to sink, for example, from a grain boundary to the surface of a material. This results in elongation of the specimen in the direction of stress, a situation found in metals, but only postulated for oxides [39]. Thus, stress at a metal–oxide interface may create vacancies in the metal surface which aid oxide growth by anion diffusion.

Balluffi and Blakely [38] have estimated that the logarithm of the diffusion coefficient D under biaxial stress varies with the stress σ_s. For $\sigma_s = 2 \times 10^9$ dyne cm^{-2}, the ratio of D (stress) to D (unstressed) is 1.24. As σ_s aproaches 10^{10} dyne cm^{-2}, the ratio is estimated to vary between 3 and 5.

Bulk oxides are classed as brittle materials. They are perfectly elastic up to the failure limit, that is, have no yield region as do metals. Hence, the model for stress effects in metals must be applied with caution to oxide films. Stress may simply open cracks and fissures in the oxide, giving rise to paths of easy migration and eventually spalling of the oxide.

Fromhold [6] has also analyzed the role of stress in oxidation. Maximum stress is found at the metal–oxide interface, relaxing to near zero at the oxide–gas interface. The stress gradient as well as the elastic strain gradient should be independent of position in the oxide. Thus, the lattice constant of the oxide perpendicular to the metal–oxide interface becomes a linear function of position as found for copper oxide on copper [40]. This stress is due to mismatch between the metal and oxide unit cells. Other sources of stress also exist. For instance, volume change due to the presence of diffusing species in the oxide can produce strain in the oxide.

While the theory of stress effects in growing oxides is developing, its application to experimental results is very limited. One of the few studies in which stress is invoked to explain oxidation at low temperatures is that of Eldridge [41]. The oxidation of lead at low temperatures is discussed in terms of an electric field generated by a piezoelectric mechanism. There is no clear-cut justification for adopting such a model, since the same data fits the Cabrera–Mott equation, as shown in Chapter 8. However, it is a novel mechanism which highlights the possible interaction of mechanical and electrical forces.

In summary, kinetic models for low-temperature oxidation are presented. Choice of a model can often be based on the structure of the oxide.

REFERENCES

1. O. Kubaschewski and B. E. Hopkins, *Oxidation of Metals and Alloys*, 2nd ed., Butterworths, London, 1962.
2. S. Mrowec and T. Werber, *Gas Corrosion of Metals*, Translated by W. Bartoszewski, National Center for Scientific, Technical and Economic Information, Warsaw, 1978 (Available from U.S. Dept. of Commerce NTIS, Springfield, VA 22161).
3. A. Paul, *Chemistry of Glasses*, Chapman and Hall, New York, 1982.
4. U. R. Evans, *The Corrosion and Oxidation of Metals: Scientific Principles and Practical Applications*, St. Martin's Press, New York, 1960, pp. 819ff.
5. C. Wagner, *Corrosion Sci.* **13** (1973), 23.
6. A. T. Fromhold, Jr. *Theory of Metal Oxidation-Vol. I-Fundamentals*, North-Holland, New York, 1976.
7. M. J. Dignam, in *Comprehensive Treatise of Electrochemistry*, Vol. 4, J. O'M. Bockris, B. E. Conway, E. Yeager, and R. E. White, Eds., Plenum Press, New York, 1981, p. 247.
8. K. Hauffe, *Oxidation of Metals*, Plenum Press, New York, 1965.
9. J. R. Anderson and I. M. Ritchie, *Proc. Roy. Soc.* (*London*) **A299** (1967), 354.
10. N. Cabrera and N. F. Mott, *Rep. Prog. Phys.* **12** (1948–49), 163; N. F. Mott, *Trans. Faraday Soc.* **43** (1947), 429.
11. A. T. Fromhold, Jr. and E. L. Cook, *J. Appl. Phys.* **38** (1967), 1546; *Phys. Rev.* **158** (1967), 600; **163** (1967), 650.
12. G. Amsel and D. Samuel, *J. Phys. Chem. Solids* **23** (1962), 1707.
13. R. B. Mosley and A. T. Fromhold, Jr., *Oxid. Metals* **8** (1974), 19, 47.
14. D. J. Young and M. J. Dignam, *J. Phys. Chem. Solids* **34** (1973), 1235.

15. F. P. Fehlner and N. F. Mott, *Oxid. Metals* **2** (1970), 59.
16. A. Many, T. Goldstein, and N. B. Grover, *Semiconductor Surfaces*, North-Holland, New York, 1965.
17. L. Young, *Anodic Oxide Films*, Academic Press, New York, 1961.
18. F. P. Fehlner, *J. Electrochem. Soc.* **119** (1972), 1723.
19. R. Ghez, *J. Chem. Phys.* **58** (1973), 1838.
20. H. H. Uhlig, *Acta Met.* **4** (1956), 541.
21. H. Uhlig, J. Pickett, and J. MacNairn, *ibid.* **7** (1959), 111.
22. V. O. Nwoko and H. H. Uhlig, *J. Electrochem. Soc.* **112** (1965), 1181 and discussion; *ibid.* **113** (1966), 639.
23. A. T. Fromhold, Jr., *Nature* **200** (1963), 1309.
24. A. T. Fromhold, Jr., *J. Electrochem. Soc.* **115** (1968), 882.
25. I. M. Ritchie, *Surface Sci.* **23** (1970), 443.
26. P. B. Needham, Jr., H. W. Leavenworth, Jr., and T. J. Driscoll, *J. Electrochem. Soc.* **120** (1973), 778.
27. L. H. Lee and V. Y. Doo, *ibid.* **118** (1971), 443.
28. D. J. Young and M. J. Dignam, *Oxid. Metals* **5** (1972), 241.
29. F. Lukeš, *Surface Sci.* **30** (1972), 91.
30. E. C. Williams and P. C. S. Hayfield, in *Vacancies and Other Point Defects in Metals and Alloys*, Institute of Metals Monograph No. 23, 1957, p. 131.
31. B. Chattopadhyay, *Thin Solid Films* **16** (1973), 117.
32. I. M. Ritchie and G. L. Hunt, *Surface Sci.* **15** (1969), 524.
33. I. M. Ritchie, *Phil. Mag.* **19** (1969), 421.
34. N. F. Mott, *Trans. Faraday Soc.* **35** (1939), 1175; **36** (1940), 472.
35. K. Hauffe and B. Ilschner, *Z. Elektrochem.* **58** (1954), 382.
36. M. J. Dignam, D. J. Young, and D. G. W. Goad, *J. Phys. Chem. Solids* **34** (1973), 1227.
37. T. B. Grimley and B. M. W. Trapnell, *Proc. Roy. Soc.* (*London*) **A234** (1956), 405.
38. R. W. Balluffi and J. M. Blakely, *Thin Solid Films* **25** (1975), 363.
39. J. S. L. Leach and P. Neufeld, *Corrosion Sci.* **9** (1969), 225.
40. B. Borie, C. J. Sparks, Jr., and J. V. Cathcart, *Acta Met.* **10** (1962), 691.
41. J. M. Eldridge, *Surface Sci.* **40** (1973), 531.

3

STRUCTURE OF VITREOUS OXIDES

3.1. INTRODUCTION

Oxide films obtained by high-temperature oxidation of metals are usually polycrystalline. In contrast, the thin, usually less than a few tens of Angstroms thick, oxide films obtained at low temperatures or by anodic oxidation are often noncrystalline. A notable exception to the above is the SiO_2 film which is noncrystalline even when the oxidation temperature is 1200°C and the oxide is relatively thick (>1 μm). The importance of noncrystalline (NC), particularly vitreous, structure in oxide films considered for passivating metals has been pointed out by Revesz and Kruger [1]. The role of noncrystallinity in the oxidation and corrosion of metals and semiconductors was discussed by Revesz and Fehlner [2]. The following discussion closely follows these two works. The important point is that the transport properties of oxide films cannot be understood without understanding their structure. However, the structure and transport properties of noncrystalline solids are much less well known than those of crystalline solids. This is the reason that the structure of noncrystalline vitreous oxides is discussed below in detail.

3.2. VITREOUS STRUCTURE

The structure of noncrystalline solids is characterized by a lack of long-range crystallographic order (LRO) which determines the structure of crystalline solids. The lack of LRO, however, is not necessarily associated with a complete lack of ordering [3]. Thus, the optical properties of silica glass are close to those of crystalline SiO_2. This behavior is due to the well-known fact that the tetrahedral configuration of four oxygen atoms surrounding each silicon atom is very nearly the same in both crystalline SiO_2 and silica glass, that is, the Si–O bond length and O–Si–O bond angle are, in the first approximation, practically independent of the structure of 4:2 coordinated SiO_2 polymorphs [4]. It has been suggested that noncrystalline solids which exhibit this type of short-range order (SRO) should be labeled vitreous in contrast to amorphous solids which have limited, if any, SRO [5].

It is obvious that this classification represents extreme cases and the transition from vitreous to amorphous is gradual. The dielectric properties in the visible and UV spectrum are reasonably sensitive indicators of SRO [6]. These properties indicate that in addition to SiO_2 glass, Ta_2O_5 can also be considered as vitreous [7], but SiO_x ($x \sim 1$) is amorphous [4,6]. In the latter material, structural disorder is associated with compositional disorder [8], whereas vitreous solids maintain the compositional order of the corresponding crystal.

The extent of ordering in vitreous solids may extend beyond the region of nearest neighbors. For example, according to Konnert et al. [7], silica glass consists of domains where the structure resembles that of crystalline tridymite. The size of these domains is about 20 Å. Phillips [9] reviews the possibility of β-cristobalite domains. This type of ordering has been labeled structured ordering as opposed to crystallographic ordering associated with LRO. Phillips [9], in particular, argues for the existence of clusters instead of continuous random networks, pointing out that noncrystalline chalcogenide films show domain structures on a 100-Å scale. The implications of these structures are not yet clear.

According to Stevels [10], structural ordering in NC solids can be expressed by the coefficient of linear repetition (CLR). This term indicates the perfection of atomic spacing along a given direction with a CLR value of unity for the perfect crystal. Vitreous SiO_2 has a CLR value of unity for a distance of 2.5 Å corresponding to the O–O distance in the $SiO_{4/2}$ unit. For greater distances, there is a reasonably high CLR value up to about 20 Å according to the model of Konnert et al., followed by a rapid decline. Thus, one can visualize that regions of various degrees of ordering are present in vitreous solids.

The lack of LRO and a reasonably high degree of SRO in vitreous solids are possible only if the chemical bonds between the atoms are flexible enough that the lack of LRO is not associated with significant reduction in the SRO through the generation of network defects. This bond flexibility may be due to relatively elastic bonds which can be sufficiently strained without breaking them, or to the ability of electrons around the atoms to rearrange themselves as the bond lengths and/or angles vary, that is, the nature of the bond varies. Since the bonds in metal oxides are usually partially covalent, they are relatively rigid, and the bond flexibility usually arises from the redistribution of bonding electrons. The bond flexibility in the latter sense results in structural flexibility. Various examples of bond and structural flexibility will be discussed below.

The vitreous state, as defined above, represents a very perfect solid in spite of the lack of LRO, comparable in perfection to a single crystal. This point is illustrated by, for instance, the facts that the difference in the standard free enthalpies of formation of the most stable and least stable forms of silica, α-quartz and vitreous SiO_2, respectively, is only about 1%, that the configurational entropy of vitreous SiO_2 is very low, 0.9 cal K^{-1} [11], and that the density of network defects in vitreous SiO_2 is usually less than 10^{16} cm^{-3} [12]. Other vitreous oxides are not as well characterized as SiO_2, but there are indications that under some conditions Ta_2O_5 can also be prepared in a highly perfect vitreous state [13].

The practical importance of vitreous oxide films lies in the fact that they can be obtained under conditions which preclude the preparation of single-crystal oxide films as, for instance, on polycrystalline substrates. Even on a single-crystal metal substrate, the formation of a vitreous oxide film does not require that the structures of the oxide and substrate be in a reasonably good epitaxial relationship, as is necessary for growing a single-crystal oxide film on a single-crystal substrate. If a vitreous or single-crystal oxide film cannot be formed, then the oxide is polycrystalline. The grain boundaries in such a solid often provide easy paths for diffusion during oxidation. As a result, they are not protective. The importance of vitreous oxide films regarding passivation of metals and semiconductors lies in their ability to uniformly slow ionic movement through the oxide, thus increasing corrosion resistance and electronic stability [1,2].

The rates and mechanisms of oxide growth are strongly affected by the vitreous structure of the oxide film. Hence low-temperature oxidation, which often results in vitreous oxides, cannot be discussed without paying close attention to the implications of the vitreous structure. However, the vitreous state is much less understood than the crystalline state.

The problem of determining under what conditions a vitreous oxide film can be formed is a complex one. It is closely related to the problem of glass formation. The terms glassy and vitreous are often used interchangeably but they can have different connotations. Glasses have historically been inorganic oxide mixtures which can be melted, formed, and annealed to form useful transparent articles. In recent times, the concept of glass has been extended to include chalcogenide, organic, semiconducting, and metallic glasses, while vitreous refers to oxide glasses.

The vitreous or glassy state is considered to be thermodynamically metastable [14], as shown in Figure 3.1. It exists because the kinetics of

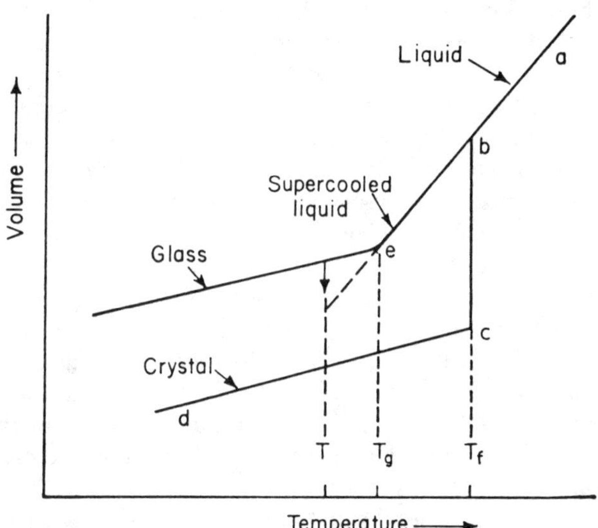

Figure 3.1. Phase diagram for relationships between the glass, liquid, and crystalline states. T_f is the crystallization temperature in the presence of nuclei while T_g is the glass-transition temperature. At temperature T below T_g, the glass volume decreases to achieve a new phase boundary. After Rawson [14]. Published by permission of the copyright owner, Academic Press, Inc.

crystallization are slow enough that a molten liquid can be thermally quenched to the solid vitreous state before nucleation and growth of crystals can occur. This has been discussed by Turnbull and Polk [15], who point out that there are additional ways to make a glass other than quenching a liquid. Splat cooling [16] or thin film techniques such as vacuum deposition [17], thermal oxidation, and anodization [18] can also be used to make vitreous materials from the gaseous or solid rather than the liquid state [19].

In general, a vitreous oxide has a network structure in which metal atoms are bound to oxygen atoms by predominantly covalent bonds. This network may be modified by the introduction of metal oxides called network modifiers which do not enter the network. Instead, they lead to the formation of nonbridging oxygens which are negatively charged, the network modifier metal being present as a positive ion. Such defects as nonbridging oxygens and dangling bonds are illustrated in Figure 3.2.

The addition of a sufficient quantity of modifying oxide to a continuous random network (CRN) causes the network to break up into structural subunits, for example, sheets, filaments, and eventually individual anion

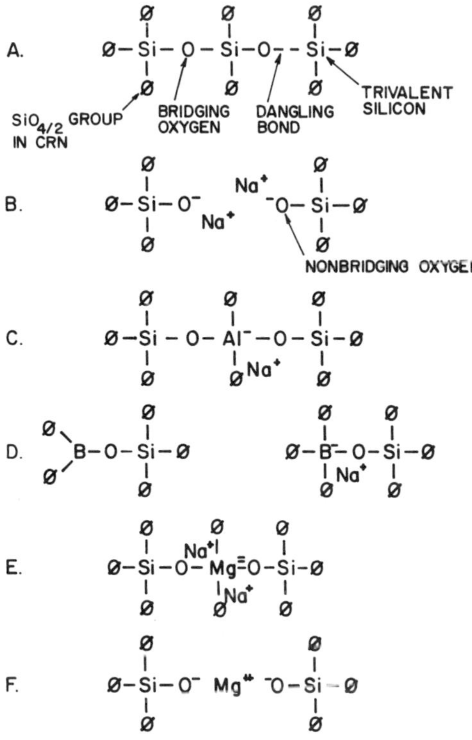

Figure 3.2. Models of bonding in vitreous oxides: (A) $SiO_{4/2}$ tetrahedra showing a broken bond; (B) Na_2O breaking up (modifying) a $SiO_{4/2}$ network; (C) an Al ion substituting for Si in fourfold coordination; (D) three- and four-coordinate boron; note the charge compensating cation in both the Al and B cases; (E) Mg substituting for Si in the network; (F) Mg acting as a modifier.

groups. The substructure is indicated by the oxygen–silicon ratio which is used to characterize silicate glasses containing more than one oxide. It gives some idea of the connectivity of the network [20].

The probability of a glass being formed by quenching a melt has been related to several physical parameters. Goldschmidt [21] utilized the ratio of cation radius to anion radius. A value of the ratio between 0.2 and 0.4 generally gives rise to a vitreous oxide. This structural concept was extended by Zachariasen [22] and Warren [4] to include irregularly interconnected building blocks such as $[SiO_4]$. Such building blocks are formed by arranging the oxygen atoms around a central cation as follows: MO and M_2O_5 form networks if the oxygen forms tetrahedra around the metal atoms; M_2O_3 forms a network if the oxygen forms triangles around the

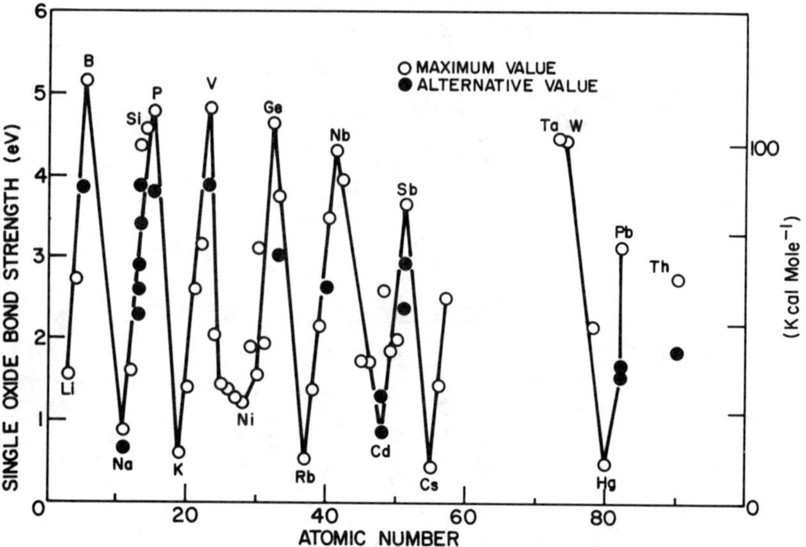

Figure 3.3. Single oxide bond strength versus atomic number of the cation. After Revesz and Fehlner [2]. Published by permission of the copyright holder, Plenum Publishing Corp.

metal; MO_3, M_2O_7, and MO_4 are probable network formers; while MO and M_2O should not form networks at all, but generally act as modifiers.

When tetrahedra and triangles form easily, a compound tends to quench to a glass [23]. Zachariasen's rules for such glass formation, as summarized by Vogel [24], are as follows. Any two such polyhedra can have only one corner in common. In other words, the anions linking the polyhedra can bond to no more than two central cations in simple glasses. Each polyhedra must have less than six corners, and at least three of these corners must link with adjoining polyhedra.

The ease of glass formation has been related to the strength of a single oxide bond by Sun and Huggins [25]. Values of bond strength as given in Table 5.2 are compared in Figure 3.3 where the periodicity with atomic number of the cation is shown [2]. Those oxides with bond strengths greater than 3.25 eV (75 kcal mole $^{-1}$) are expected to be glass formers; those with bond strengths between 2.18 and 3.25 eV are called intermediates which means that they may either enter a glassy network or act as a modifier of an existing network; and those with bond strengths less than 2.18 eV (50 kcal mole $^{-1}$) are modifiers and generally crystallize when formed at room temperature or above.

Dietzel [26] has shown that the concept of field strength of the ion, F'

$= Z/y^2$, where Z is the cation valence and y is the interval between ions, leads to the same division of elements into network formers and modifiers as the bond energies in Figure 3.3. This concept provides an explanation for the unexpected formation of crystals in binary melts of two network-forming oxides. The oxygen ions tend to surround that cation which has the higher field strength. This leaves the other (larger) cation to bond with the resulting complex; for example, Na^+ with $[SiO_4]^{4-}$. In such a case, crystallization of compounds can occur with $\Delta F' > 0.3$ (Å)$^{-2}$. Even though the Zachariasen rules are fulfilled, crystals form in the SiO_2–P_2O_5 and B_2O_3–P_2O_5 systems for high concentrations of P_2O_5. On the other hand, SiO_2–B_2O_3 remains vitreous over the whole span of composition [24]. In like manner, aluminate, nitrate, and carbonate glasses can be accounted for.

The ideas of Dietzel have been extended by Smekal [27], who stressed the need for mixed ionic and covalent bonds in glasses. Weyl [28] discussed the role of screening in the stability of complex ions and polymerization of the coordination groups which make up a glass; for example, $[SiO_4]$, $[BO_3]$, $[PO_4]$, or $[BeF_4]$. Thus, the glass network becomes a statistical distribution of modifying cations spread over a randomly ordered, polymerlike network. These ideas are closely related to those of bond flexibility discussed above. This concept will be covered in greater detail below.

The microcrystalline theory of Lebedev [29] has continued to be discussed and modified to bring it closer in concept to the CRN theory. However, it still remains basically different in that definite crystalline linkages or even crystalline assemblages are assumed to exist in glass. Phillips [9] has presented some related ideas. He holds for a cluster model in which partially broken chemical order stabilizes internal interfaces. Nonbridging oxygens are thought to exist at the cluster surfaces, stabilizing cluster dimensions of 25–35 Å for silica.

Two points are of interest in the above classifications. In general, the same oxides should form glasses in all cases. Secondly, potentially vitreous oxides which do not form glasses by quenching from the melt can be converted into vitreous oxides by film techniques.

It should be emphasized that thermodynamics and related concepts, for example, bond strength, are not sufficient to explain the formation of vitreous oxides. The reason is that kinetic considerations may be of overriding importance. The kinetic theory of glass formation from a melt, as advanced by Turnbull and Cohen [30], is considered by Rawson [14] to be more comprehensive than the Zachariasen theory. In the kinetic theory, the irregular network structure of a quenched melt requires bond breakage to transform to a regular crystalline structure. The free energies of activation for this reconstruction process should be high in relation to the temperature at which crystals can be formed if the vitreous oxide is to be stabilized.

However, in the present consideration of vitreous oxide grown directly from a metal, other factors dominate. For instance, impurities such as water can determine the nature of the oxide. Water can promote vitreous oxide by forming multiple structures of equivalent energy, or destroy vitreous oxide by weakening network bonding, allowing crystallization. Thus, structure and bond energy are emphasized to highlight the role of vitreous oxides in protecting metals from oxidation and corrosion and in stabilizing oxide–semiconductor interfaces.

It must be stressed that the vitreous oxide network is a special type of structure, akin, for example, to siloxane polymers, and is not characteristic of all inorganic noncrystalline materials. For instance, modifying oxides may be found as noncrystalline materials if formed at sufficiently low temperatures. However, instead of a network structure, they will be more like solid argon or glassy metals: a collection of hard spheres frozen into liquidlike disorder. Such a solid is then better described as amorphous rather than vitreous. It is interesting to note that network-forming oxides can be vitreous, crystalline, or amorphous, while modifiers tend to be crystalline or amorphous.

3.3. RELATIONSHIP OF VITREOUS OXIDE TO OXIDATION

It is critical to the oxidation process that self-diffusion and/or permeation take place for continued oxide growth. Self-diffusion occurs in single crystals or ideal vitreous oxides containing no paths for easy ion movement. Permeation takes place through channels in vitreous oxide or along defect lines or planes such as grain boundaries in polycrystalline oxide. For example, a comparison of oxygen diffusion rates in aluminum oxide [31] shows that the presence of grain boundaries increases the measured diffusion rate due to permeation of oxygen along the grain boundaries.

Pawel and Campbell [32] have also shown that the noncrystalline structure of an anodic tantalum oxide film on tantalum can slow the rate of oxygen permeation and, therefore, the rate of oxide growth.

The concept of a channel in a vitreous oxide needs to be defined at this point. Two types of channels can exist. The first is statistical in nature and arises from a correlation of the five- and six-membered rings observed in models of fused silica [33].

A second type of channel is postulated to form due to the anisotropic movement of ions through a growing oxide. These channels are taken to be more or less perpendicular to the substrate, as in SiO_2 growing on

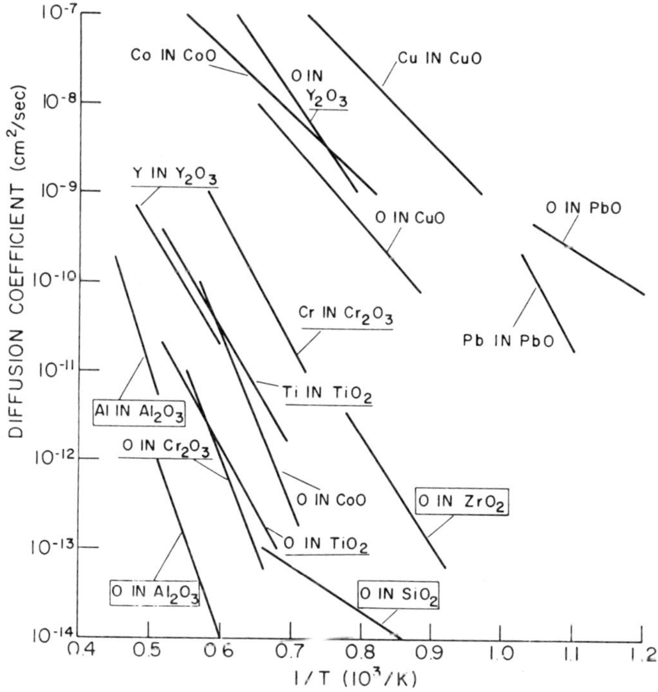

Figure 3.4. Logarithm of the self-diffusion coefficients for ions in oxides as a function of reciprocal temperature. Network formers are noted by boxes and intermediates by underlining. The other examples are modifiers. After Revesz and Fehlner [2]. Published by permission of the copyright holder, Plenum Publishing Corp.

silicon [34]. They gradually merge into the bulk vitreous oxide structure as the oxide thickens.

The application of the above discussion to low-temperature oxidation or corrosion is as follows. When ionic rather than electronic movement in the oxide controls the rate of reaction (the usual case), the faster moving ion, anion or cation, will determine the rate of oxidation. Figure 3.4 shows that the rate of self-diffusion [35] of the faster ion in a particular oxide is generally slower in crystalline network-forming oxides than in crystalline network-modifying oxides. For example, the rate of growth of Al_2O_3 on Al, which is controlled by Al ion movement, can be compared with the rate of growth of CoO on Co where Co ion movement is rate controlling. It is obvious that the Co ions in CoO move faster than the Al ions in Al_2O_3, and hence the rate of oxide growth on a Co single crystal will be greater.

Crystallization of an oxide also relates directly to ease of ion movement. Hence, network-forming oxides tend to be vitreous, intermediate oxides occur in both vitreous and crystalline form, while modifiers are polycrystalline at room temperature. As a result, the effective diffusion rate for ions in the polycrystalline modifying oxides on metals is expected to be equal to or greater than that shown in Figure 3.4 for single crystal oxides.

Single crystal oxide films are seldom encountered in practice, while vitreous oxides are often found on metals. Since both types of oxide have the same short-range order and lack the gross defects which encourage fast ion movement, then by analogy, the conclusion of Figure 3.4 for single crystal oxides should be applicable to vitreous oxides. This figure illustrates the advantage of vitreous network-forming oxides in limiting oxidative attack on metals. Oxide growth should be slowest if the oxide is vitreous or single crystal, and fastest if polycrystalline. An added advantage of the vitreous state is the low density of dangling bonds which can promote ion movement in oxide films.

3.4. CRYSTALLIZATION OF VITREOUS OXIDE

Since the vitreous state is thermodynamically less stable than the crystalline one, it is possible for crystallization to occur. The driving force may be either temperature [36], electric field, or stress. Oxidation can be quite dependent on the nucleation and growth of crystallites from the vitreous oxide. This is so because ion motion through the oxide film is often rate determining and recrystallization of a vitreous oxide provides paths of easy-ion movement at the grain boundaries.

Nucleation can proceed by one of two processes. The first, homogeneous nucleation, occurs in the bulk of a glass, sometimes following phase separation or precipitation of a supersaturated solute. The second, heterogeneous nucleation, is difficult to avoid, for it takes place on container walls, at imperfections, or on impurity particles. The mathematics of both processes are treated extensively by Kingery et al. [37a]. Experimentally, the maximum supercooling achieved for materials which are not network formers is ~20% of the melting point [38]. This limit has been exceeded for network formers, splat-cooled glassy metals, and thin films deposited on cold substrates.

Crystal growth proceeds after the nuclei have formed. Two models have been proposed to account for the growth process. In "rough" growth, atoms can be added or removed from any site on the crystal surface. The growth rate can be expressed as

$$u_1 = \nu' a_0 \left[1 - \exp\left(-\frac{\Delta G}{kT} \right) \right] \tag{1}$$

TABLE 3.1. ESTIMATED COOLING RATES FOR GLASS FORMATION[a]

Material	dT/dt (°K sec^{-1}), Homogeneous nucleus, $\Delta G^* = 50\ kT$, $T_r = 0.2$	dT/dt (°K sec^{-1}), Heterogeneous nucleus, $\beta = 80°$, $\Delta G^* = 50kT$, $T_r = 0.2$	dT/dt (°K sec^{-1}), Homogeneous nucleus, $\Delta G^* = 60\ kT$, $T_r = 0.2$
$Na_2O{\cdot}2SiO_2$	4.8	46	0.6
GeO_2	1.2	4.3	0.2
SiO_2	7×10^{-4}	6×10^{-3}	9×10^{-5}
Salol	14	220	1.7
Metal	1×10^{10}	2×10^{10}	2×10^{9}
H_2O	1×10^{7}	3×10^{7}	2×10^{6}

[a] In the headings, ΔG^* is the free energy for nucleation, T_r is the relative undercooling of the melt, and β is the contact angle between the nucleus and the substrate. After Kingery et al [37b]. Published by permission of the copyright holder, John Wiley & Sons, Inc.

where u_1 = growth rate per unit area
ν' = frequency factor for transport of ions to the growing crystal surface
a_0 = distance advanced by interface in unit kinetic process
ΔG = free energy change (proportional to the undercooling)

$$\nu' = \frac{kT}{3\pi a_0^3 \eta} \tag{2}$$

where η is the viscosity. The limiting growth rate is estimated to be 10^5 cm sec^{-1}. A large number of step sites must be available on the crystal surface for this model to apply and generation of such steps can require a high activation energy. For this reason, a second model of crystal growth has been advanced based on screw dislocations.

Screw dislocation growth does not require the nucleation of growth steps on the crystal surface, and thus little or no activation [39]. Instead, the screw dislocation provides a self-perpetuating source of steps. The rate u_2 is

$$u_2 = fu_1 \tag{3}$$

where f is the fraction of preferred growth sites at dislocation ledges.

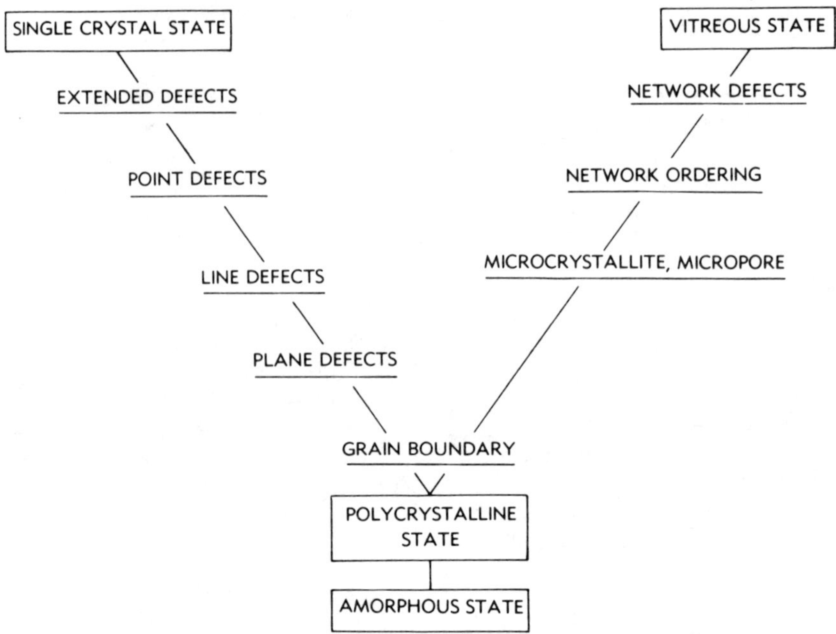

Figure 3.5. Tentative classification of defects in vitreous oxides in comparison with those in crystalline oxides. Increasing defect magnitude leads to polycrystalline oxide and finally amorphous oxide (no long- or short-range order). After Revesz and Fehlner [2]. Published by permission of the copyright holder, Plenum Publishing Corp.

This brief discussion of crystal nucleation and growth can be applied to glass formation by calculating time–temperature–transformation plots [40]. Assuming that 10^{-6} is the smallest fraction of crystallized material which can be detected, undercooling versus time can be plotted for the transformation of glass to crystallites, as for example, $Na_2O{\cdot}2SiO_2$ [41]. Estimated cooling rates needed for glass formation, as taken from Kingery [37b], are given in Table 3.1. Such data are extremely useful for materials which crystallize readily.

It is clear from the above discussion that, regarding transport processes, the state of greatest imperfection is a polycrystalline body with grain boundaries that allow easy ion movement. With respect to the overall order, the amorphous state is the least ordered, but it is not clear that this affects the transport properties when compared to the vitreous state. There are different degrees of imperfection between the extremes as illustrated in Figure 3.5. The nature of defects in crystalline materials and their influence on the course of oxidation are well documented [42]. Although the same is not yet true for vitreous oxides, some attempts to define these conceptually different defects have been initiated. Revesz [43] has suggested that some defects in vitreous SiO_2 are unique for the vitreous state. For example, a structural channel which is a part of perfection in a crystal can arise in vitreous SiO_2 from a helical arrangement of Si and O atoms, resembling those occurring in tridymite and α-quartz. In the vitreous oxide, such channels are defects. Likewise, growth of areas of perfection in a vitreous oxide represents more extended defects. These result in microcrystallites which can grow to form large crystals embedded in a glassy matrix, that is, a glass ceramic. Complete crystallization of the glassy phase produces a polycrystalline body.

In a related fashion, a single crystal can transform to a polycrystalline body through the formation and growth of defects. Point defects such as vacancies can grow into line and plane defects which coalesce into grain boundaries. Even the amorphous state can, through reordering the short-range structure and growth of crystallites, achieve a polycrystalline structure.

3.5. ROLE OF BOND AND STRUCTURAL FLEXIBILITY

The formation and stability of vitreous oxides, particularly SiO_2 and Ta_2O_5, are closely related to structural and/or bond flexibility as exemplified by organic or metal–organic polymers and polymorphs of elemental solids [44]. The concepts of structural and bond flexibility will be illustrated for SiO_2 and Ta_2O_5, as well as for other oxides which are important in modern technology.

3.5.1. Oxide Films on Silicon, Germanium, and Chromium

Vitreous SiO_2 is formed during the oxidation of silicon single crystals which are the most perfect and best understood solid-state material available. Owing to its great importance in semiconductor device technology, the Si/SiO_2 interface has been extensively studied. An important factor in these investigations has been the preparation of the noncrystalline SiO_2 films on a "perfect" (chemically pure, single crystal) substrate under very clean conditions. As a result, the SiO_2 films are also "perfect," that is, vitreous, although several factors such as high temperature (up to ~1300°C), single crystal substrate, and slow growth rate (less than 0.5 Å sec^{-1} as the average rate for 1000-Å oxide growing at 1000°C) favor the formation of polycrystalline SiO_2 films. These films are usually vitreous even when thicker than 1 μm.

The electronic properties of the silicon substrate are very sensitive indicators of the perfection of an SiO_2 film, including the Si/SiO_2 interface.

These indicators are unsurpassed by any other physical–chemical analytical technique. Consequently, the Si/SiO_2 interface has emerged as the best characterized solid–solid interface. The perfection of thermally grown SiO_2 films can be illustrated by the low density of sodium impurity ($<10^{15}$ cm^{-3}) and electron trapping sites ($<10^{14}$ cm^{-3}), high dielectric breakdown strength ($>10^6$ V cm^{-1}), as well as excellent reproducibility and stability under extreme conditions of temperature and electric field. The perfection of the Si/SiO_2 interface is demonstrated by the very low density of interface states: $<10^{10}$ cm^{-2} corresponding to $\sim 10^{-5}$ monolayers of adsorbate. These interface states are associated with either chemically active, unsaturated (dangling) bonds of surface Si atoms which, at an ideal Si/SiO_2 interface, should be bonded to oxygen atoms, and/or with electron (hole) trapping defects in the oxide film within tunneling distance ($\sim$30 Å) of the interface.

A high perfection of the SiO_2 film and of the Si/SiO_2 interface is due to the noncrystalline structure of SiO_2. When the oxide begins to crystallize, for example, during prolonged heat treatment at high temperatures or in the presence of contamination, the electrical properties of the film deteriorate and the density of interface states increases. The first effect results from the disruption of the Si–O network as crystallites form. This process is associated with the generation of network defects and later even with the formation of grain boundaries between crystallites. As mentioned above, the presence of grain boundaries in the oxide film can lead to a decrease in its passivating (protective) character. Thus, a polycrystalline or highly defective noncrystalline oxide film is a poor candidate for passivation, either chemical or electronic.

The increase in interface state density during crystallization of the SiO_2 film results from the loss of structural flexibility that is characteristic of the SiO_2 film while it is noncrystalline. Flexibility of the vitreous oxide allows the majority of the surface Si atoms to bond to an oxygen atom of the oxide without requiring an almost perfect epitaxial relationship between Si and SiO_2. However, when the SiO_2 film is crystalline, the match of two structures having rigidly defined but different long-range order is poor and hence the interface is disordered. Apparently, quasiperfect solid–solid interfaces can only be obtained when a quasiperfect epitaxial relationship exists between two single crystals, or at least one side of the interface is noncrystalline. This conclusion is very pertinent to the passivation of metals which are usually (but not exclusively) polycrystalline and, therefore, an epitaxial relationship with the oxide film is restricted to some individual crystallites in the polycrystalline oxide rather than the film as a whole. Thus, elimination or drastic reduction of the active sites in the thin passive film requires that the passivating film be noncrystalline.

The perfection of noncrystalline oxide relies on the fact that the lack of LRO is not caused by total destruction of structural order. Rather, the SRO characteristic of crystalline polymorphs of SiO_2 is maintained in noncrystalline SiO_2 films as well as in bulk vitreous silica; that is, every Si atom is surrounded tetrahedrally by four O atoms and the Si–O bond length is constant. It is the Si–O–Si bridging bond angle which is characterized by a wide distribution (from ~120° to ~180°) in vitreous SiO_2 [45]. It should be noted that the distribution of the Si–O–Si bond angle is quite large, not only in vitreous SiO_2, but in crystalline tridymite as well, ranging from ~140° to ~173° in the unit cell comprising 320 SiO_2 units [46]. As mentioned above, there are regions in vitreous SiO_2 extending to ~20 Å which resemble tridymite or cristobalite. The wide distribution of the Si–O–Si bond angle makes it possible to establish a Si–O network without long-range order and without Si–O bond breaking which would introduce a large number of defects. This type of noncrystalline structure is, of course, vitreous in contrast with amorphous noncrystalline structures which do not have appreciable short-range order. (Fused silica and SiO_2 films on silicon are vitreous solids *par excellence*, whereas, for instance, vacuum-deposited SiO is an amorphous solid.) The lack of short-range order in amorphous solids is usually associated with chemical disorder, whereas the short-range order in vitreous solids maintains stoichiometry and ensures reproducibility.

The structural flexibility of SiO_2 is manifested in the large number (nine) of crystalline polymorphs which have essentially identical short-range order and almost the same standard free enthalpy of formation for which the difference between quartz and fused silica is about 1%. Nevertheless, they exhibit significant variations in properties such as density, compressibility, and chemical reactivity, because various topological arrangements of $SiO_{4/2}$ tetrahedra can be present in these polymorphs. Thus, similar to organic and siloxane polymers, various arrangements of atoms are, in the first approximation, independent of energy, resulting in a variation in structure which is essentially conformational. This point is illustrated in Figure 3.6, where the experimental distributions of the Si–O–Si bridging bond angle ϕ' in crystalline SiO_2, as well as the calculated variation in the bridging bond energy as a function of ϕ', are shown. The value of $\Delta\mathscr{E}$ reaches a minimum at the point where bond angle reaches a preferred value. The variation in the bridging bond energy was calculated by a molecular orbital method for the pyrosilicic acid molecule $(OH)_3SiOSi(OH)_3$, which contains an Si–O–Si bridging bond [47]. Owing

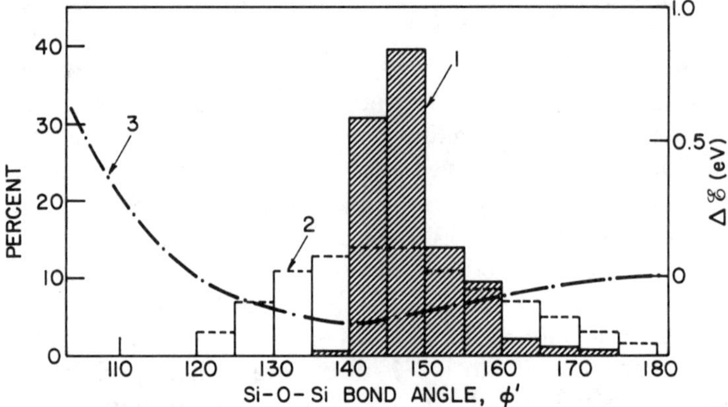

Figure 3.6. The differential energy $\Delta\mathscr{E}$ of $H_6Si_2O_7$ in curve 3 compared with the distribution of Si–O–Si bridging bond angles ϕ' in tridymite crystals (bar graph 1) and vitreous silica (bar graph 2). Reprinted with permission from *Proc. Conf. Phys. MOS Insulators*, G. Lucovsky et al., Eds., p. 92, A. G. Revesz and G. V. Gibbs, Structural and Bond Flexibility of Vitreous Silicon Dioxide Films, Copyright 1980, Pergamon Press, Ltd.

to the localized nature of the Si–O bonds, many properties of SiO_2 can be illustrated by the acid.

The reason for the behavior shown in Figure 3.6 is that the total Si–O bond overlap population, particularly its π component, increases by 3.6% and 4.7% respectively as ϕ increases from 140° to 180°. Concomitant with this variation, the Si–O bond length decreases from 1.615 to 1.594 Å. The mean value at the peak of the ϕ distribution is 1.60 Å. The latter value is practically identical with the experimentally determined mean value for vitreous silica, 1.60 Å [45]. It should be noted that the average negative charge on oxygen in vitreous SiO_2 is calculated to be 0.56*e*, indicating that the Si–O bond is largely covalent rather than ionic. In the latter case the charge would be 2.0*e*.

The very small variation of the energy with the Si–O–Si bridging bond angle helps explain the fact that thermal oxidation of silicon results in a vitreous SiO_2 film which has unique stability.

Despite the close resemblance of germanium to silicon, the properties of GeO_2 are rather different from those of SiO_2 even though the Ge–O bond is also a mixed ionic–covalent bond, basically in the same proportion as the Si–O bond. However, there is no π-bonding in GeO_2 because of much less overlap between the oxygen 2*p* and germanium 4*d* orbitals relative to the Si 3*d* orbitals. As a result, the bond flexibility is greatly reduced. There is only one GeO_2 crystalline polymorph with 4:2 coordination (hexagonal, silica structure) and the tendency toward noncrystallinity is much less than in the case of SiO_2. The lack of π-bonding in GeO_2 is also partially responsible for the significant solubility of hexagonal, noncrystalline GeO_2 in water. Hence, the thin GeO_2 film formed at room temperature does not passivate to the extent that the SiO_2 film does. This difference in the passivation behavior is the main reason that silicon has replaced germanium in semiconductor device technology.

A different kind of bond flexibility is exhibited by metals in various oxidation states. For example, the standard free enthalpies per gram atom of Cr are -6.12 and -5.91 eV for CrO_3 and Cr_2O_3, respectively. This difference is relatively small so that various structural conformations of approximately the same energy may exist, encouraging the formation of relatively stable noncrystalline oxides. In addition, transition metal oxides such as Cr_2O_3 may exhibit a departure from the regular structure due to ligand effects [48]. Since both σ-bonding and π-bonding can be involved in these effects, and π-bonding can affect the stereochemistry while its contribution to the bond energy may be relatively small, a situation somewhat similar to that characteristic of the Si–O bond may arise. Even without involving π-bonding, ligand effects may increase bond flexibility. It is possible that these effects are partially responsible for the important role of H_2O in passivation since H_2O is a σ-donor ligand. While the effects of H_2O will be discussed below, it is significant to mention in this context that, due to the large number of d electrons, chromium (electron configuration in ground state: [Ar] $3d^5 4s^1$) can be associated with water as a ligand in several configurations. This may be an additional factor in increasing the stability of the noncrystalline passivating film on Cr metal and in the beneficial effects of Cr alloying on the corrosion resistance of Fe alloys. This behavior does not exclude the possibility that a polycrystalline oxide film forms on top of the thin vitreous oxide film on Cr in such a manner that the structure of the vitreous oxide is preserved. The net result is that Cr may exhibit a significant resistance to oxidation under these conditions.

3.5.2. Structural Complexity and Vitreous Structures—Tantalum and Aluminum Oxides

It is well known that anodic oxidation of tantalum results in a noncrystalline Ta_2O_5 film whose dielectric and optical properties, as well as its excellent reproducibility, indicate that there is a high degree of short-range order in the structure; that is, the film is vitreous. Crystallization studies of stripped anodic oxide films show that a heat treatment of 600°C/30 min in vacuum is necessary to initiate crystal growth [36] although anodic films may be crystallized under field at less than 100°C [49]. On the other hand, thermal oxidation of tantalum at ~500°C, both in polycrystalline and single crystalline form, usually results in polycrystalline oxide with poorly defined properties [13]. Apparently, the thin noncrystalline passivating film on the Ta surface crystallizes above some critical temperature, reducing its protective nature. In contrast, it grows thicker without crystallization during anodic oxidation at room temperature. However, it has been found [13] that vitreous Ta_2O_5 film can be obtained by thermal oxidation of a microcrystalline Ta film vacuum deposited on silicon. This Ta_2O_5 film is so morphologically and optically perfect that it is used as an antireflection film in silicon solar cells of improved conversion efficiency. The vitreous Ta_2O_5 film on silicon can withstand prolonged heat treatments up to 500°C without crystallization, perhaps due to slight Si incorporation in the oxide. Polycrystalline Ta foil oxidized under similar conditions results in a polycrystalline Ta_2O_5 film with poor optical properties. This behavior is an example of the effect of the substrate upon the structure of the oxide film, to be discussed below.

The structure of Ta_2O_5 must be considered to explain the occurrence of the vitreous state. According to Stephenson and Roth [50], the structure of Ta_2O_5 is based on chains of distorted octahedral and pentagonal bipyramidal oxygen polyhedra around the Ta atoms. The coordination of oxygen in these configurations is not 2. Hence, simple Ta–O–Ta bridges analogous to Si–O–Si bridges do not exist and the Zachariasen rules of glass formation are unfulfilled. However, Ta_2O_5 is a network former on the basis of metal–oxygen bond strength.

Since the Ta–O bond is ~60% ionic and ~40% covalent, the bond is expected to have some flexibility. Additional bond flexibility may result from ligand field effects characteristic of transition-metal oxides. This is probably the reason for the large variation (from 1.83 to 2.60 Å) of the Ta–O bond length in the low-temperature polymorph of crystalline Ta_2O_5. (Compare this variation with that of the Si–O–Si bond angle in tridymite discussed above.) The structural complexity of Ta_2O_5 is manifested in the very large unit cell that is divided into subcells [51]. The arrangement of these subcells and the size of the unit cell can depend on heat treatment and impurities such as hydrogen and silicon. For example, the unit cell formula at temperatures below 1000°C is $Ta_{28}O_{70}$.

The structural complexity is closely associated with the possibility of varying the composition (stoichiometry) by changing the relative numbers of subunits which are structurally closely related but differ slightly in composition. These structures have been named "infinitely adaptive" [52] or "vernier" [53] structures. Since the difference in lattice energy between the alternative structural arrangements is very low, the various substructures can easily interchange. In this regard, the structure of Ta_2O_5

exhibits a similarity to organic polymers and analogously, the various arrangements of substructures may be considered as structural conformations.

It is very likely that the adaptive nature of the Ta_2O_5 structure plays an important role in the formation of noncrystalline Ta_2O_5. Specifically, it is suggested that the increased tendency toward long-range order usually exhibited by structures in which the nonmetallic element has a coordination larger than 2 is counteracted to some extent by the structural complexity. As a result, arranging the atoms according to the long-range order of the crystal during the formation of the oxide may be more difficult kinetically than forming a vitreous structure.

The coordination of Al and O atoms in Al_2O_3 is 6:3, so that the oxygen does not form Al–O–Al bridges and Al_2O_3 cannot strictly be considered a network former. However, there are five polymorphs of Al_2O_3 (corundum) whose standard free enthalpies of formation are within 2.5% of each other. The structural difference among these polymorphs arises from different degrees of ordering Al ions in an essentially closest packing of oxygen ions [54]. This structural flexibility is quite different from that exhibited by SiO_2, but it could be important in the formation of noncrystalline Al_2O_3 as well as in other ionic oxide structures based on nearly closest packing of oxygen ions. Such an oxide is, for instance, the polycrystalline Fe–Cr perovskite which forms on stainless steel with low Cr content [55]. It is likely that the structural flexibility of the perovskite structure is increased by the incorporation of chromium. This could be an important reason for the film becoming more noncrystalline and, hence, more protective at higher Cr content.

It is evident from the above examples that structural/bond flexibility is an essential factor in determining whether an oxide film can be obtained in a reasonably stable vitreous form so that it can passivate the underlying metal. In addition to the structural/bond flexibility, other factors such as the presence of additives, the chemical and physical nature of thin oxide films, and the role of the metal substrate influence formation and stability during growth of a vitreous oxide film. These factors, which influence the kinetics as well as the thermodynamics of the process, are discussed below.

3.5.3. Effects of Additives to Oxide Films

Additives to the oxide film can be introduced, not only by alloying the bulk metal, but also by surface treatment of the metal, for example, by ion implantation, pack cementation, or electrolysis in fused salts. In view of the previous discussion, it is obviously desirable that the additive enhance noncrystallinity of the passivating film. This can be achieved by increasing the structural/bond flexibility, for instance, Cr in the passivating film on stainless steel. Another possibility is to increase the stability of the noncrystalline structure by decreasing the ease of crystallization, for example, incorporation of Si into Ta_2O_5 during the oxidation of Ta on silicon [13].

Special attention should be given to hydrogen as an additive in oxide films. Hydrogen in the oxide film can lead to increased structural flexibility by forming M–H or M–OH bonds in addition to the normal M–O bonds. If the corresponding metal oxyhydroxide has various polymorphs, for instance, AlO·OH [56], then various conformations of the M–OH bond may exist in the H-containing passivating film. This results in increased tendency toward noncrystallinity. The M–OH groups can be linked together via hydrogen bonds as in the case of AlO·OH and CrO·OH [56], facilitating the formation of a noncrystalline network. This is similar to hydrogen-bonded organic polymers. The fact that hydroxide ions of some metals (e.g., Cr, Al, Ta, and Si) can form hydroxyl-bonded polymers, but others (e.g., Fe, Ni, and Co) [57] cannot, is also pertinent to the structural flexibility of noncrystalline films. Several researchers have provided evidence for hydrogen or "bound water" [58] in both crystalline [59,60] and noncrystalline passive films.

On the other hand, H or OH may act as a modifying oxide and weaken the oxide structure if the water destroys a strong network such as SiO_2 [61,62]. Recrystallization of an oxide may be facilitated by a mineralizer such as water and can result in a greatly increased rate of oxidation [60,63] due to grain boundary formation. This is similar to the role of NaCl or other ionic impurities in destroying the structure of a protective oxide.

Another important property of hydrogen (water) in passivating films is its role in self-repair, as pointed out by Okamoto [58]. This effect is somewhat analogous to the reduction of interface states (dangling Si bonds) at the Si/SiO_2 interface by forming Si–H or Si–OH bonds [64] or the action of corrosion inhibitors [65]. In this manner, the reactive sites can be tied up so that the apparent perfection of the film and interface is increased. However, the Si/SiO_2 interface which contains hydrogen is more sensitive to destructive processes, for example, irradiation, than a quasiperfect Si/SiO_2 interface in which practically all Si atoms are bonded to oxygen.

Hydrogen may be present even in films grown at high temperatures under relatively clean conditions. For example, SiO_2 films on silicon may contain SiOH and/or SiH groups up to $\sim 10^{21}$ cm^{-3}, particularly at the Si/SiO_2 interface [66]. Evidently, a very high concentration of hydrogen is to be expected in thin passivating films formed in room air or in aqueous environments. In such cases, hydrogen should be considered as a major constituent of these films rather than an additive. This is an important area for detailed studies designed to gain better insight into the complex behavior of hydrogen in passivation.

3.5.4. Substrate Effects

Rate control of oxidation may shift in some instances from movement in the bulk oxide to reaction at an interface. In any case, ion incorporation at the metal–oxide interface depends on the perfection of the metal or semiconductor surface, that is, the number of kink and ledge sites available for easy-ion incorporation and the chemical homogeneity of the metal.

Substrate effects in silicon and tantalum oxidation have been pointed out above. Substrate effects may also be evidenced by a decrease in the rate of reaction of amorphous metals as compared to polycrystalline alloys of the same composition [67]. For instance, the corrosion resistance of an iron-based amorphous metal has been ascribed to chemical homogeneity and the lack of inclusions [68]. Eldridge and Dong [69] point out that the absence of grain boundaries and dislocations on a metal surface can be important. Devine and Wells [70] have observed that, although no pitting, that is, breakdown of the passivating film, occurred on amorphous metals, it did occur after the metals were crystallized by heat treatment. A study of fast-quenched $Fe_{1-x}B_x$ metallic glass likewise showed an increase in corrosion current with crystallization [71]. Reviews of the chemical properties of amorphous metals have been published [72,73].

Laser glazing is a method used to convert the surface of a metal to a noncrystalline state. This process also slows the rate of corrosion, for example, Al [74].

Fan and Henrich [75] report that a dense amorphous film of germanium oxidized in dry air at room temperature formed an oxide much more slowly than a polycrystalline film of germanium. They also noted that a porous, amorphous film of germanium allowed oxygen penetration into the interior of the film, resulting in oxidation rates similar to those for the polycrystalline germanium.

Ion implantation offers a method for both doping a metal surface and converting it to an amorphous state [76,77,78]. Novel alloys can be formed in this manner [79] and the corrosion process itself can be studied [80]. Work by Townsend et al. [81] summarizes the efforts of several investigators concerning the effect of ion implantation on oxidation and corrosion. Two major processes have been considered: radiation damage and impurity doping. It is expected that, even when the main purpose of ion implantation is to introduce an additive which improves passivation behavior, effects due to disordering of the metal surface will arise.

The enhanced corrosion resistance of many ion-implanted crystalline metals resembles that of amorphous metals. This behavior may be due at least partially to the increased stability of noncrystalline (vitreous) oxide films on the damaged (in many cases probably amorphous) metal layer resulting from ion implantation.

The formation and stability of a noncrystalline film are determined by the balance between the competing processes of oxide growth and crystallization. Crystal nucleation rate in the noncrystalline oxide film, and hence the stability of the noncrystalline structure, depend heavily on the substrate, both chemically and physically.

3.5.5. Thin-Film Effects

An important, yet frequently overlooked, point is the fact that passivating films are very thin. The usual considerations derived from bulk behavior are not necessarily applicable to thin-film phenomena. For example, PtSi can be formed at ~200°C by an interface reaction between the Si substrate and thin (several hundred angstroms thick) Pt film, whereas according to the phase diagram, the minimum temperature should be above 700°C [82]. Also, an oxidizing reaction due to oxygen diffusion through Ta_2O_5 takes place at a Si/Ta_2O_5 interface at a rate which is higher than that on bare silicon at the same temperature [13]. Kinetic rather than thermodynamic considerations are crucial in thin-film phenomena where the interface energy is often high and overcomes other energy barriers.

These thin-film effects are further complicated by the structure of noncrystalline oxide films. Unfortunately, structural information obtained from X-ray or electron diffraction measurements is insufficient to characterize these films. There is a great need for applying techniques which give information on short-range order and on the nature of the chemical bonds to complement electrochemical measurements.

3.5.6. Ductility

It was pointed out above that oxides are usually rigid and, hence, considerable stress may be present in oxide films on metals. However, there are some indications that the degree of crystallinity affects film ductility. Bubar and Vermilyea [83] showed that anodic Ta_2O_5 lost its ductility after crystallization. Bubar and Vermilyea [84] have also shown that there is a wide variation in the ductility of the films formed on a number of metals. Those formed on Ta, which is known to form noncrystalline films, exhibited deformations of as much as 50% before fracture. Also, the passive film on 304 stainless steel exhibited better ductility than that on crystalline Fe. Aluminum, which may or may not form crystalline films, formed films less ductile than those on Fe. Leach and Neufeld [85], however, found easy deformation of anodic oxide films on Al under an applied field. Arnott et al. [86] showed that fracture of anodic oxide on aluminum depends on the defect density in the oxide. Cracks in anodic oxide have been detected by acoustic emission [87]. Plastic flow of anodic films on tantalum has also been demonstrated [88]. A caution is needed here though, since too high a field can result in film crystallization [18,49].

In summary then, it is clear that vitreous oxides can be a key element in controlling reactivity of a metal. Examples based on Si, Ta, Al, Ge, and Cr support this conclusion. However, many unanswered questions remain.

REFERENCES

1. A. G. Revesz and J. Kruger, in *Passivity of Metals*, R. P. Frankenthal and J. Kruger, Eds., The Electrochemical Society, Princeton, NJ, 1978, p. 137.
2. A. G. Revesz and F. P. Fehlner, *Oxid. Metals* **15** (1981), 297.
3. R. Zallen, *The Physics of Amorphous Solids*, John Wiley & Sons, New York, 1983.
4. B. E. Warren, *Z. Kristallogr.* **86** (1933), 349.
5. A. G. Revesz, *Phys. Stat. Sol.* **19** (1967), 193.
6. S. H. Wemple, *J. Chem. Phys.* **67** (1977), 2151.
7. J. H. Konnert, J. Karle, and G. A. Ferguson, *Science* **179** (1973), 177.
8. J. Pivot, D. Morelli, J. A. Roger, and C. H. S. Dupuy, *Thin Solid Films* **34** (1976), 205.
9. J. C. Phillips, *Solid State Phys.* **37** (1982), 93.
10. J. M. Stevels, *J. Non-Cryst. Solids* **6** (1971), 307.
11. R. J. Bell and P. Dean, *Phil. Mag.* **25** (1972), 1381.
12. D. L. Griscom, in *Defects and Their Structure in Nonmetallic Solids*, B. Henderson and A. E. Hughes, Eds., Plenum Press, New York, 1976, p. 323, and references therein.
13. A. G. Revesz and T. Kirkendall, *J. Electrochem. Soc.* **123** (1976), 1514 and references therein.
14. H. Rawson, *Inorganic Glass-Forming Systems*, Academic Press, New York, 1967.
15. D. Turnbull and D. E. Polk, *J. Non-Cryst. Solids* **8–10** (1972), 19.
16. H. Jones, *Rep. Prog. Phys.* **36** (1973), 1425.
17. L. Holland, *Vacuum Deposition of Thin Films*, Chapman and Hall, London, 1966.
18. L. Young, *Anodic Oxide Films*, Academic Press, New York, 1961.
19. D. R. Uhlmann, *J. Non-Cryst. Solids* **25** (1977), 43.
20. J. M. Stevels, in *Encyclopedia of Physics*, Vol. 20, S. Flügge, Ed., Springer-Verlag, Berlin, 1957, p. 350.
21. V. M. Goldschmidt, *Geochemische Verteilungsgesetze der Elemente*, Skrifter Norske Videnskaps. Oslo, Mat.-Nat. Kl. No. 7, 1926, p. 7.
22. W. H. Zachariasen, *J. Amer. Chem. Soc.* **54** (1932), 3841–57.
23. A. Paul, *Chemistry of Glasses*, Chapman and Hall, New York, 1982.
24. W. Vogel, *Structure and Crystallization of Glasses*, Pergamon Press, New York, 1971, pp. 14–18.
25. K-H. Sun and M. L. Huggins, *J. Phys. Colloid Chem.* **51** (1947), 438.
26. A. Dietzel, *Z. Elektrochem.* **48** (1942), 9; *Glastech. Ber.* **22** (1949), 212.
27. A. Smekal, *Glastech. Ber.* **22** (1949), 278.
28. W. A. Weyl, *Silicates Ind.* **23** (1958), 309, 374, 458, 518, 587, 662; **24** (1959), 36, 88, 151, 208, 267, 321.

29. A. A. Lebedev, *Arb. Staatl Opt. Inst. Leningrad* **2** (1921) No. 10, 1–20; See also I. Schulz and W. Hinz, *Silikattechnik* **6** (1955), 235.
30. D. Turnbull and M. H. Cohen, in *Modern Aspects of the Vitreous State*, vol. I, J. D. Mackenzie, Ed., Butterworths, Washington, 1960, p. 38.
31. Y. Oishi and W. D. Kingery, *J. Chem. Phys.* **33** (1960), 905.
32. R. E. Pawel and J. J. Campbell, *J. Electrochem. Soc.* **113** (1966), 1204.
33. D. L. Evans and S. V. King, *Nature* **212** (1966), 1353.
34. A. G. Revesz and H. A. Schaeffer, *J. Electrochem. Soc.* **129** (1982), 357.
35. Per Kofstad, *Nonstoichiometry, Diffusion, and Electrical Conductivity in Binary Metal Oxides*, Wiley-Interscience, New York, 1972.
36. R. E. Pawel and J. J. Campbell, *J. Electrochem. Soc.* **111** (1964), 1230.
37. W. D. Kingery, H. K. Bowen, and D. R. Uhlmann, *Introduction to Ceramics*, 2nd ed., John Wiley & Sons, New York, 1976; (a) p. 328ff; (b) p. 350; (c) p. 260.
38. K. A. Jackson, *Nucleation Phenomena*, American Chemical Society, Washington, DC, 1965.
39. A. Van Hook, *Crystallization*, Reinhold, New York, 1961.
40. R. A. Grange and J. M. Kiefer, *Trans. Am. Soc. Metals* **29** (1941), 85.
41. G. S. Meiling and D. R. Uhlmann, *Phys. Chem. Glasses* **8** (1967), 62.
42. F. A. Kröger, *The Chemistry of Imperfect Crystals*, Vol. 3, 2nd ed., American–Elsevier, New York, 1974.
43. A. G. Revesz, *J. Non-Cryst. Solids* **4** (1970), 347.
44. R. Wang and M. D. Merz, *Nature* **260** (1976), 35.
45. R. L. Mozzi and B. E. Warren, *J. Appl. Cryst.* **2** (1969), 164.
46. J. H. Konnert and D. E. Appleman, *Acta Cryst.* **B34** (1978), 391.
47. A. G. Revesz and G. V. Gibbs, in *MOS Insulators, Proc. Int. Top. Conf.*, G. Lucovsky et al., Eds., Pergamon Press, New York, 1980, p. 92.
48. C. S. G. Phillips and R. J. P. Williams, *Inorganic Chemistry*, Vol. 1, p. 480 and Vol. 2, pp. 189–239, Oxford University Press, New York, 1965.
49. D. A. Vermilyea, *J. Electrochem. Soc.* **102** (1955), 207; **104** (1957), 542.
50. N. C. Stephenson and R. S. Roth, *Acta Cryst.* **B27** (1971), 1037.
51. R. S. Roth and J. L. Waring, *J. Res. NBS* **74A** (1970), 485.
52. J. S. Anderson, *J. Chem. Soc.(London)* **10** (1973), 1107.
53. B. G. Hyde, A. N. Bagshaw, S. Andersson, and M. O'Keeffe, *Ann. Rev. Mater. Sci.* **4** (1974), 43.
54. A. F. Wells, *Structural Inorganic Chemistry*, 4th ed., Clarendon Press, Oxford, 1975, p. 459.
55. C. L. McBee and J. Kruger, *Electrochem. Acta* **17** (1972), 1337.
56. A. F. Wells, *op. cit.*, pp. 526–529.
57. C. S. G. Phillips and R. J. P. Williams, *op. cit.*, **1**, p. 533.
58. G. Okamoto, *Corrosion Sci.* **13** (1973), 471.
59. H. T. Yolken, J. Kruger, and J. P. Calvert, *ibid.* **8** (1968), 103.
60. C. L. Foley, J. Kruger, and C. J. Bechtoldt, *J. Electrochem. Soc.* **114** (1967), 994.
61. F. P. Fehlner and N. F. Mott, *Oxid. Metals* **2** (1970), 59.
62. A. G. Revesz and R. J. Evans, in *Reactivity of Solids* (*Proc. Int. Symp., 6th, 1968*), J. W. Mitchell et al., Eds., John Wiley & Sons, New York, 1969, p. 425.
63. P. B. Sewell, D. F. Mitchell, and M. Cohen, *Surface Sci.* **29** (1972), 173.
64. A. G. Revesz, *J. Non-Cryst. Solids* **11** (1973), 309.
65. H. H. Uhlig, *The Corrosion Handbook*, John Wiley & Sons, New York, 1948.
66. K. H. Beckmann and N. J. Harrick, *J. Electrochem. Soc.* **118** (1971), 614.
67. *Metallic Glasses* (ASM Seminar, Sept. 1976), American Society for Metals, Metals Park, OH, 1978.

68. T. M. Devine, *J. Electrochem. Soc.* **124** (1977), 38.
69. J. M. Eldridge and D. W. Dong, *Surface Sci.* **40** (1973), 512.
70. T. M. Devine and L. Wells, *Scripta Met.* **10** (1976), 309.
71. P. Kovács, J. Farkas, L. Takács, M. Z. Awad, A. Vértes, L. Kiss, and A. Lovas, *J. Electrochem. Soc.* **129** (1982), 695.
72. T. Masumoto and K. Hashitomo, *Ann. Rev. Mater. Sci.* **8** (1978), 215.
73. R. B. Diegle, N. R. Sorensen, T. Tsuru, and R. M. Latanision, in *Corrosion: Aqueous Processes and Passive Films*, J. C. Scully, Ed., Academic Press, New York, 1983, p. 59.
74. P. L. Bonora, M. Bassoli, G. Cerisola, P. L. De Anna, G. Battaglin, G. Della Mea, and P. Mazzoldi, *Thin Solid Films* **81** (1981), 339.
75. J. C. C. Fan and V. E. Henrich, *Appl. Phys. Lett.* **25** (1974), 401.
76. G. Dearnaley, J. R. Morris, and R. A. Collins, in *Proc. 7th Intern. Vac. Cong. and 3rd Intern. Conf. Solid Surfaces,* R. Dobrozemsky et al., Eds., Vienna, 1977, p. 955.
77. G. Dearnaley, in *New Uses of Ion Accelerators*, J. F. Ziegler, Ed., Plenum Press, New York, 1975, p. 295.
78. J. M. Poate, *J. Vac. Sci. Technol.* **14** (1977), 529.
79. V. Ashworth, R. P. M. Procter, and W. A. Grant, *Thin Solid Films* **73** (1980), 179.
80. J. K. Hirvonen, *J. Vac. Sci. Technol.* **15** (1978), 1662.
81. P. D. Townsend, J. C. Kelley, and N. E. W. Hartley, *Ion Implantation, Sputtering and Their Applications*, Academic Press, New York, 1976, p. 262.
82. J. W. Mayer and K. N. Tu, *J. Vac. Sci. Technol.* **11** (1974), 86.
83. S. F. Bubar and D. A. Vermilyea, *J. Electrochem. Soc.* **114** (1967), 882.
84. S. F. Bubar and D. A. Vermilyea, *ibid.* **113** (1966), 892.
85. J. S. L. Leach and P. Neufeld, *Corrosion Sci.* **9** (1969), 225.
86. D. R. Arnott, W. J. Baxter, and S. R. Rouze, *J. Electrochem. Soc.* **128,** (1981), 843.
87. J. T. Dickinson, D. B. Snyder, and E. E. Donaldson, *Thin Solid Films* **72** (1980), 223.
88. M. Propp and L. Young, *J. Electrochem. Soc.* **126** (1979), 624.

4

CHARGE TRANSPORT IN NONCRYSTALLINE OXIDES

4.1. INTRODUCTION

When a compact layer of oxide is formed on a metal, the rate of subsequent oxidation is usually assumed to depend on the rate at which oxygen or metal or both can move through it. Since these species can in principle be charged or uncharged, electronic motion is also of interest. For crystalline oxides, there is little reason to doubt that ionic transport occurs either through the formation of a vacancy or of an interstitial, and that electrons and holes move in conduction and valence bands. A pair of ion vacancies of opposite sign is called a Schottky defect; an interstitial metal ion together with a metal vacancy is a Frenkel defect. But we can also consider, for instance, an interstitial metal ion together with an electron in the conduction band, or in anodic oxidation simply an interstitial ion or a metal ion vacancy. Thus, Cabrera and Mott [1] envisaged that in anodic oxidation metal ions moved from kink sites on the metal, renewed for instance by the Frank–Read spiral process, into an interstitial site in the oxide. The cation moved across the oxide to be neutralized and form additional oxide at the oxide–electrolyte interface. Charge compensation for this process was supplied by the external circuit. However, there is no external circuit in thermal oxidation so that charge compensation must take place by electronic transport through the oxide.

4.2. ELECTRONIC CONDUCTION

The injection of electronic species into and transport across vitreous oxides is primarily a function of oxide thickness and temperature, with the interfacial energy barrier as a variable [2]. See Figure 2.6. The current–voltage relationships often encountered are summarized below.

Oxides less than ~30 Å thick can support electron tunneling at room temperature under high fields. An electron in the metal has a finite probability of existing outside the boundaries of the metal up to a distance of say 30 Å, thus penetrating the oxide barrier and supplying the charge balance needed to maintain an ionic current. The electronic current [3a] follows the relationship

$$J_e = \frac{\phi}{4\pi^2 \hbar x^2} \exp\left[\frac{-x(8m'\phi)^{1/2}}{\hbar}\right] \tag{1}$$

where J is the current density, ϕ is the barrier to electronic movement, m' is the effective mass of the electron, and $\hbar$ is Planck's constant divided by 2π.

The interaction of tunneling electrons with the oxide is the basis for tunneling spectroscopy [4]. Here, a spectrum, analogous to an infrared spectrum, is produced by the interaction of the tunneling electrons with molecules absorbed in the oxide.

Electrons may tunnel into the conduction band of thicker oxides which are subjected to a high field, or to electron traps which soon fill up, impeding further injection of charge. In this case, the current becomes

$$J = \frac{bE^3}{\phi} \exp\left[\frac{b'\phi^{3/2}}{E}\right] \tag{2}$$

for $V > \phi$ where E is the electric field and b and b' are constants.

Thermionic or Schottky emission becomes the principle mode of electron injection for thicker oxides at higher temperatures. In this case, thermal energy is available to activate electrons over the interfacial energy barrier. The Richardson–Schottky relationship controls the current [2]:

$$J = A'' \exp\left\{\frac{-[\phi - 3.8 \times 10^{-4}(E/\kappa)^{1/2}]}{kT}\right\} \tag{3}$$

where κ is the dielectric constant of the oxide, T is the absolute temperature, A'' is the thermionic emission constant, and k is the Boltzmann constant. The field is established by ionic diffusion resulting from a gradient of the ionic species across the oxide [3b].

If the current due to tunnel or thermionic injection exceeds the ability of the oxide to transport the charge, a virtual cathode forms and space-charge-limited current ensues. Space-charge-limited current in the presence of traps has also been analyzed.

The behavior of electrons in noncrystalline materials has been discussed by Mott and Davis [5]. Impurity conduction in which electrons tunnel from one impurity site to another can account for electron transfer over distances of at least several hundred Angstroms at low temperatures. For low fields, an analysis of impurity conduction leads to a modified Poole–Frenkel equation. See below.

For large fields, impurity conduction is replaced by transport in the conduction band, electrons being removed from donors according to the Poole–Frenkel equation. Thus, conduction relies on the field ionization of traps located in the band gap of the oxide. The electrons in the conduction band produce a current which follows a relationship very similar to Equation 3 for Schottky emission. The major difference is a factor of 2 in the exponential.

$$J = J_0 \exp \left\{ \frac{-[\phi_i - 7.6 \times 10^{-4}(E/\kappa)^{1/2}]}{kT} \right\} \tag{4}$$

where ϕ_i is the energy required to ionize an impurity within the oxide and J_0 is a constant.

The physics of electronic conduction in nonoxide glasses has been studied in detail during the past two decades. Its importance for the subject matter of this book is perhaps mainly the information that it has given about the defects to be expected in a continuous random network (CRN).

Much of the experimental work has been on the amorphous semiconductors, particularly deposited films of silicon and germanium and films of chalcogenide glasses, for example, As_2Te_3, either deposited or quenched from the melt. Some of the main conclusions are as follows [5].

Conduction bands and valence bands exist as in crystals, and measurements of the thermopower show whether a material is *p* type or *n* type. Due to disorder, the mean free path of a carrier is often very small and can be comparable with the electron wavelength. Under these conditions, the Hall effect does not give a reliable measure of the sign of the carrier, and it is doubtful whether a complete theory yet exists. In amorphous materials, the states near the extremities of a band are believed to become localized, that is, they turn into shallow traps. The range of energies in which these states lie forms a "tail" to the band concerned, shown for a conduction band in Figure 4.1. The range $\Delta\mathscr{E}$ of energies where states are localized is separated from that in which they are extended by an energy $\mathscr{E}_c$ called the mobility edge. Calculations of $\Delta\mathscr{E}$ for the continuous random network of amorphous silicon have been carried out by Davies [6] who obtained a value of 0.3 eV. Experimental work on drift mobility of electrons establishes a $\Delta\mathscr{E}$ value of 0.2 eV for electrons in silicon deposited by glow discharge from silane and 0.1 eV for holes in

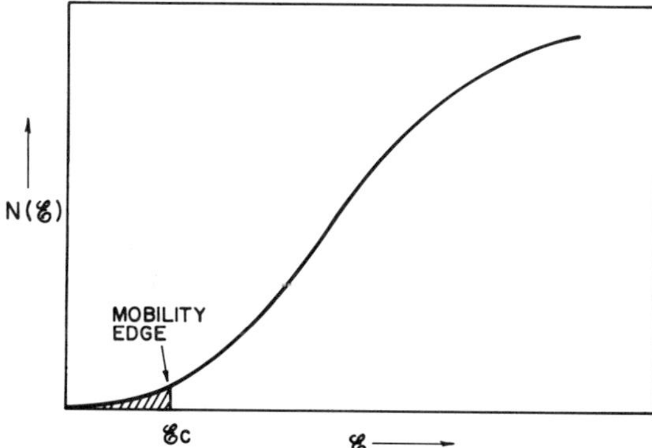

Figure 4.1. Density of states function in an amorphous semiconductor.

As_2Te_3. On the other hand, in vitreous SiO_2, $\Delta\mathscr{E}$ for electrons must be less than kT as the drift mobility decreases with increasing temperature [7]. There is also the possibility that a carrier, particularly a hole, forms a polaron giving an activated mobility not essentially associated with disorder. Whether this occurs in chalcogenides is controversial, but it almost certainly does for holes in SiO_2 [8].

For vitreous oxides, therefore, we conjecture that a conduction band with comparatively high mobility exists ($\sim$20 cm^2 V^{-1} sec^{-1} for SiO_2), but that the mobility for holes is smaller. The upper part of the valence band is formed from the oxygen lone pair orbitals, which are nonbonding so that the band is narrow. This will lead either to a value of $\Delta\mathscr{E}$ probably larger than kT, or in addition, to polaron formation. In either case the mobility is activated, in the latter case with an activation energy of $\sim$0.4 eV for holes in SiO_2.

A very striking result for the chalcogenide semiconductors obtained from the extensive work of Kolomiets [9] and coworkers is that these materials apparently cannot be doped. The activation energy for conduction is near to half the optical band gap, as for an intrinsic semiconductor, and is little dependent on composition. Thus, the addition of Si to As_2Te_3 does not give rise to donors, as it would if Si replaced As in a crystalline lattice. In fact, the crystallization of an amorphous semiconductor may give rise to an enormous increase in conductivity, the impurities acting as donors or acceptors only in the crystalline state. The explanation is that in a glass the stable configuration is one in which each atom has a coordination number which allows all electrons to be taken up in bonds. Thus germanium would have coordination number 4, arsenic

3, tellurium 2. There is considerable experimental evidence for this hypothesis on glasses. In deposited films there are certainly exceptions. Thus Spear and coworkers [10] have obtained doped amorphous silicon by codepositing in a glow discharge SiH_4 and PH_3. Much of the phosphorus does appear to go in with threefold coordination so that it is electrically inactive, but some goes in with the fourfold coordination of the silicon, forming donors. The presence of hydrogen is essential to take up dangling Si bonds which form trap sites and dominate conductivity. For vitreous oxides, we suspect that most impurities that are network formers do conform to the complete bonding rule, though whether this is necessarily so for oxide formed by low-temperature or anodic oxidation is uncertain.

4.2.1. States in the Gap

Whether it is legitimate to speak of "point defects" in a glass has proved to be controversial [11]. It is argued that since a glass is noncrystalline, a defect in a noncrystalline structure is not to be distinguished from a statistical fluctuation in bond length or angle. Such an argument is, we believe, correct for a liquid metal or metallic glass. But for oxide or chalcogenide glasses and for amorphous silicon and germanium, we think that a valid model starts with a CRN where all atoms have the same coordination number as in the crystal, and we then introduce defects [12] of which the negative dangling bond is the most discussed, though doubtless there are many others, for example, Si–Si bonds in SiO, trivalent silicon (E' centers), and peroxide bonds. (See Figure 3.2.) The dangling bond is familiar in oxide glasses under the name "nonbridging oxygen." Such a center carries a free spin on a lone pair orbital when neutral, but it is usually positively or negatively charged. In silicon and germanium, such centers are believed to exist, either as point defects or at the surface of voids, giving an electron spin resonance signal. Their concentration is greatly reduced by the presence of hydrogen or fluorine. They also form deep donor and acceptor states in the gap. The energies will be broadened by disorder into two "bands." If the two bands overlap, some of the centers will be charged. The result is that the Fermi energy $\mathscr{E}_F$ is pinned, and a new form of conduction can occur, in which an electron with energy near $\mathscr{E}_F$ "hops" from one center to another. This form of conduction leads to a conductivity following the law [13]

$$\sigma = A_4 \exp\left(-\frac{B_4}{T^{1/4}}\right) \tag{5}$$

which in evaporated silicon and germanium seems to be the predominating

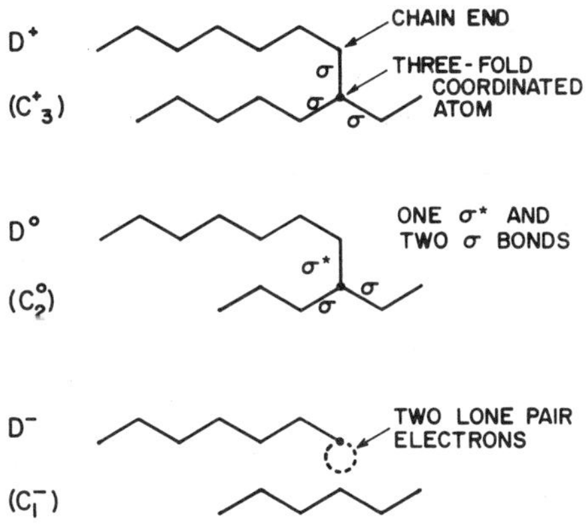

Figure 4.2. Illustration of valence alternating pair formation in selenium.

form of conduction below room temperature. It does not occur (or is greatly weakened) in specimens containing hydrogen.

It is not normally observed either in the semiconducting chalcogenide glasses, and also in these, there are no paramagnetic centers. It might then be thought that these materials contained no defects, were it not for the extensive evidence that for them too, the Fermi energy is pinned near midgap. For many years this was a puzzle, but an acceptable explanation arose from a concept due to Anderson [14], the "negative Hubbard U," which was adapted to the concept of "dangling bonds" by Street and Mott [15] and Mott, Davis, and Street [16]. Additional insights are provided by Kastner, Adler, and Fritzsche [17]. We think the theory is applicable to oxide glasses [18,19] and indeed to all noncrystalline materials in which the upper part of the valence band is formed from lone pair oxygen or chalcogen orbitals. This excludes Si and Ge. We think then that the theory may be of some importance for understanding the formation of noncrystalline oxides, and we give an outline of it here.

We start with the familiar concept of a negative nonbridging chalcogen, denoted by Street and Mott by D^- (D for defect) and by Kastner et al. by C_1 (C for chalcogenide), so as to denote that it is coordinated to one arsenic or silicon. If an electron is removed from it, leaving a singly occupied lone pair orbital and a free spin, the idea is that it can form a bond with the lone pair orbital of an adjacent chalcogen, so that the two atoms move toward each other. The bond then contains two bonding σ orbitals and one σ* antibonding orbital. For Se, the simplest case, the configuration is as in Figure 4.2. If another electron (the σ*) is removed, the bonding becomes much stronger and the two chalcogens move closer together. The atom toward which the chain-end moves is now threefold coordinated (D^+ or C_3) as pointed out by Kastner et al., positively charged, and without spin. The assumption is now made that, given two "dangling bonds," at a distance from each other so that they do not interact, the reaction to form a valence alternating pair,

$$2D^0 \rightarrow D^+ + D^- \tag{6}$$

is exothermic because the energy released in forming D^+ is so great that it compensates for the term $\langle e^2/r_{12} \rangle$ incurred by putting two electrons on the same lone pair orbital in D^-.

If this is correct, then any "dangling bonds" formed in a deposited film, or in one grown anodically, for instance, will be either positively or negatively charged and carry no spin. In glasses quenched from the melt, it is often and correctly stated that dangling bonds are unlikely to form because their energy of formation is too great in comparison with kT. But Kastner et al. point out that, to form a charged pair D^+ and D^- (a valence alternating pair in their notation), one does not have to permanently break a bond, because D^- forms one bond and D^+ three, the total number being the same as before.

In putting forward this hypothesis of charged centers, Street and Mott referred to the experimental evidence obtained earlier by Street, Searle, and Austin [20] that photoluminescence occurs when radiation is absorbed at specific defect centers, that there is a large Stokes shift, and that these centers are charged. The model was used to explain many phenomena, particularly as providing charged traps which limit the drift mobility of electrons and holes and act as recombination centers. If a material containing these centers were lightly doped, say with donors, they would lose their electrons to form an excess of D^- over D^+. The energy needed to form a free electron in the conduction band by the reaction

$$D^- \rightarrow 2 \text{ electrons} + D^+ \tag{7}$$

is not altered, and this is what we mean by saying that the Fermi energy is pinned.

The next question is the extent to which these concepts can be applied to oxide glasses. All the evidence presently available relates to SiO_2 and to some extent to soda glasses. First, it seems that a free hole forms a polaron and moves with an activated mobility [7,8]. Mott [19] has suggested that the mechanism is that described here; namely, that an oxygen on which the hole sits forms a bond with the lone pair $2p$ orbital and the two move together. Also the work of Mott and coworkers [21] on the silicon inversion layer gives strong evidence that thermally grown SiO_2

contains positively and negatively charged centers, and that these can be identified with the surface states of silicon technology. If so, the hypothesis has been made that these are charged dangling bonds and that the reaction [6] is exothermic. Pepper [22] has shown that most of them can be removed by annealing. In fact, in annealed vitreous silicon dioxide the concentration of any charged center appears very low, as one can deduce from the high mobility of injected electrons.

In pure vitreous SiO_2 one can envisage the following centers,

1. A low concentration of "charged" dangling bonds.
2. Oxygen vacancies. Here two silicon atoms form a pair when the center is neutral. It acquires a spin when positively charged, and is then called an E' center.
3. An Si–Si bond, present in silicon-rich material.

Among impurities, Na_2O is, of course, important. When incorporated in a glass, Na^+ is a network modifier, sitting in an "interstitial" position and compensated by a negative nonbridging oxygen. Other additions to SiO_2 are possible. Phosphorus replacing silicon in the network is positive and again compensated by nonbridging oxygen. Boron can be three- or fourfold coordinated. When Na^+ is present, boron is fourfold coordinated and negative. It does not seem possible to compensate it by D^+. Hydrogen (or water) is incorporated as OH^- in the network, each H_2O giving rise to two hydroxyls bound to metal atoms as $M\text{–}O^-H^+$. The hydroxyl is thought to be neutral with no spin, and in comparison with O^{2-} is a positively charged center, compensated by D^- (nonbridging oxygens).

4.3. IONIC MOVEMENT IN VITREOUS OXIDES

A knowledge of the movement of atoms and ions in a network structure is one key to understanding, on an atomic basis, the process of oxide growth at low temperatures. In crystals, lattice ion movement is considered to take place by one of four mechanisms [23].

Vacancy. Movement of an ion on a normal site into an adjacent unoccupied lattice site.

Interstitial. Movement of an ion from one interstitial site to an adjacent interstitial site.

Interstitialcy. Movement of an interstitial ion onto a lattice site, thereby displacing the occupying ion to an interstitial position (knock-on).

Free Transport. Movement of an ion, atom, or molecule down a channel or grain boundary.

The first two mechanisms are well known for crystalline materials. Interstitialcy is observed when energetic particles bombard a sample. It has also been proposed as a mechanism for ion movement in glass. The fourth mechanism is proposed for silicon oxidation. A fifth mechanism, ring diffusion, has also been proposed for crystalline materials [24]. It is similar to place exchange [25], postulated to take place in oxide films, but it has been considered less probable in bulk oxides. It must be borne in mind, however, that diffusion in oxide films can differ from that in bulk oxides due to the presence of large area free surfaces, short-circuit diffusion paths, biaxial stress, impurities, and steep chemical and/or electrical gradients [26,27]. Thus, place exchange is also a possible mechanism for ion movement during oxidation of metals as discussed below.

Structural data as well as kinetic constants can be derived from diffusion, that is, permeability, studies in vitreous materials. Permeability is defined as $D \times S$ for small values of S where D is the diffusion coefficient and S the solubility. For movement of a noble gas through glass as shown in Figure 4.3, permeability has been related to the mole percent of network former present in the glass [28] and is a function of temperature as expressed by the Arrhenius equation [29]. The explanation for Figure 4.3, in which permeation rate decreased as percent network former decreased, has been given in terms of modifying cations blocking holes in the network structure of the vitreous oxide. A magnitude for the hole radius in fused silica has been estimated to be 0.3 Å based on the activation energies for diffusion of gases with different radii [30]. A similar behavior is noted in the mixed alkali effect [31] where diffusion of one alkali ion in an oxide network is impeded by the presence of a second alkali ion [32]. A similar mixed halide effect has also been found in glasses where anions are mobile [33].

4.3.1. Diffusion of Network-Modifying Ions

In bulk glass, the motion of modifying ions in a network structure has been examined in detail. Stevels [34] summarizes experimental work on modifying cations by pointing out two limiting factors; geometry and charge density. Some recent calculations of Huggins [35] strengthen this viewpoint. If a channel or interstitial region is small, then large ions cannot move easily and smaller ions will have the larger mobility. On the other hand, if the channel is large, then the geometric factor is no longer limiting. Instead, the larger ion, that is, the one with a smaller charge density, will have the higher mobility because it will be bound less tightly to adjoining ions. In a particular network, a balance between ion size and charge density leads to an optimum size for maximum diffusion rate, for example,

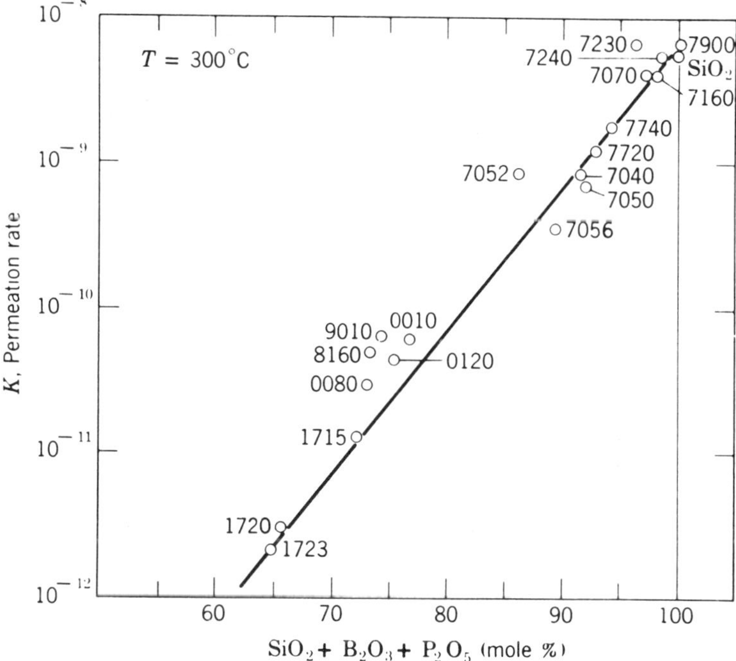

Figure 4.3. Variation of the permeation rate of helium with the concentration of network formers (SiO_2, B_2O_3, P_2O_5) in glass. After Altemose [28]. Published by permission of the copyright holder, the American Institute of Physics.

Na^+ and Ag^+ in silicate glasses. Mott [19] discusses the nature of sodium ions in SiO_2, their movement, and interaction with nonbridging oxygens.

The movement of modifying ions ionically bonded to the oxide network can be likened to the interstitial motion of ions in alkali halide crystals [24,36]. Hence, the process is at least partly understood. Divalent cations move much more slowly than monovalent cations and tend to block movement of the monovalent species if both are present.

It is interesting to note that the diffusion of gases changes very little in the region of the glass transition temperature, indicating that a liquid film of vitreous oxide could still retain some protective character during oxidation. However, modifying cations show a distinct increase in diffusion rate as the glass passes through the glass transition temperature.

Increases in the percent alkali in a network former such as silica lead to an increasing rate of diffusion for the alkali, presumably because the network is being broken up. However, in a glass where change of coordination number is possible, for example, in borates, the addition of alkali strengthens the network initially, although leading eventually to breakup of the network. Rapid quenching of glasses also leads to increased diffusion rates, implying a more open network structure.

4.3.2. Diffusion of Network-Forming Ions

Self-diffusion of ions in crystalline oxides has been reviewed by Kofstad [23] and Seltzer [37]. However, the movement of network-forming ions through a vitreous oxide is less well understood than the movement of modifiers. Either movement of charged defects incorporated in the oxide or movement of network-forming species through the established network may occur during oxidation. The former process might be expected to correlate with the viscosity of glass [31]. As pointed out by Dignam [38], defects can arise in high-field conductivity by injection at an interface or by homogeneous pair generation within the oxide.

There is evidence from studies of oxygen diffusion in fused silica [29] and of silicon oxidation [39] at high temperatures without an applied field that molecular diffusion takes place through the oxide. This is discussed more fully in the treatment of silicon oxidation (Chapter 9). Molecular diffusion at high temperatures through vitreous oxides other than SiO_2 has yet to be reported. In fact, during anodic oxidation of tantalum, silicon, aluminum, and other vitreous-oxide-forming metals, ions are known to move under the influence of an electric field.

Some of the best information on the movement of network-forming ions comes from anodic thin film studies. It must be kept in mind that these studies are concerned with specific oxides in a liquid or gaseous electrolyte under the influence of an electric field, rather than with all oxides. Nevertheless, a general picture of self-diffusion for network ion motion in a vitreous oxide has begun to emerge.

Dell'Oca, Pulfrey, and Young [40] have summarized various models of ion movement in anodic oxides. These models are unsatisfactory in terms of explaining all experimental observations, but they are useful in terms of giving direction to further experimental work.

Early ideas involved cation and anion lattices slipping through each other under steady-state conditions. Either interface or bulk control of the process was possible. As models developed, transient effects observed during current–voltage studies led to the idea of field-created Frenkel or Schottky defects. An ionic avalanche was a modification of this theory. Dignam [41], on the other hand, has proposed a slow polarization model to explain constant field transients.

It has been found that the logarithmic dependence of anodic current density on applied field E is not strictly linear. The replacement of βE in

Equation 23 of Chapter 2 with $-\gamma' E^{1/2}$ or $-\alpha' E + \beta' E^2$, where α', β', γ' are constants, more precisely expressed the results. The $\beta' E^2$ term describes the nonlinear Tafel slope, $\Delta(\log J)/\Delta E$. Models to explain the E^2 or $E^{1/2}$ dependence have been based on electrostriction, changes in activation distance, and variations in barrier potential.

The $-\gamma' E^{1/2}$ term is discussed by Young and Zobel [42] in terms of a vitreous oxide. Metal ions are assumed to travel freely down meandering channels in the oxide but become trapped at occasional excesses of negative charge. For noninteracting coulombic centers of charge, the analogy to the Poole–Frenkel law for electronic currents results, such that

$$J = J_0 \exp\left(-\frac{W_t - \gamma' E^{1/2}}{kT}\right) \tag{8}$$

where W_t is the trap energy. Further development of these ideas has included structural changes in the oxide [43] and the existence of a two-layer oxide. A general review of the subject has been presented by Dignam [44].

Revesz and Fehlner [45] have classified the defects which can be found in a vitreous oxide, thereby facilitating ion movement through the oxide. Their classification includes a comparison with crystalline defects and is shown in Figure 3.5. Of particular note is the fact that disorder in crystalline materials results in defects, while just the opposite is true in vitreous oxides. Here, ordering of the ions produces a defect. It is important to recall that we are considering topologically disordered, noncrystalline solids, that is, vitreous oxides having short-range order but no long-range order and characterized by a bridging bond angle having a distribution about a mean value. We see from Figure 3.2 that a broken network bond, also called a nonbridging oxygen or dangling bond, upsets the perfect short-range order.

A different result is found if there is a nonrandom distribution of bridging bond angles. When the distribution is uncorrelated, then microheterogeneities, similar to the five- and six-membered rings of the Evans–King model of SiO_2 occur [46]. More importantly, if the distribution is correlated, then channels based on a statistical correlation of rings could be formed. These channels would provide paths of easy-ion and even molecular movement. Channels may also form as a direct result of the growth process. It is, of course, possible for channels and dangling bonds to occur simultaneously, enhancing the ease of ion movement still further, as discussed above for the model of Young and Zobel.

A model for self-diffusion in network-forming oxides can be constructed based on the postulate of channels and dangling bonds. Stevels'

TABLE 4.1. IONIC TRANSPORT NUMBERS FOR OXIDES

Metal Group	Oxide[a]	Cation	Anion	Reference
I	Cu_2O	~1	~0	86, 87
II	BeO	0.75	0.25	88
III	Al_2O_3 [b]	0.6, 0.4	0.4, 0.6	70, 75
IV	ZrO_2 [c]	<0.05	>0.95	75
	HfO_2 [c]	<0.05	>0.95	75
	SiO_2	<0.05	>0.95	72
V	V_2O_5	0.28	0.72	89
	Nb_2O_5	0.27	0.73	75
	Ta_2O_5	0.24	0.76	69
VI	WO_3	0.37	0.63	75

[a] Anodic oxides except Cu_2O which was formed by thermal oxidation.
[b] Transport number varies with current density and electrolyte used in anodization.
[c] Oxide was microcrystalline as formed.

work suggests that large ions can travel more easily through channels than through a close-packed array of atoms. It has been suggested [47] that the movement of larger anions would predominate through the open structure of vitreous, network-forming oxides while movement of smaller cations would preferentially occur through the more compact, normally polycrystalline but more loosely bound modifying oxides. Movement of both species might be expected in an intermediate oxide. Such a result has been found for anodic oxides as shown in Table 4.1 where the sum of ionic transport numbers has been taken to be one.

The place exchange mechanism offers an alternative explanation for self-diffusion in vitreous oxides [48,49]. It has been extended to account for simultaneous cation and anion [25] movement. In place exchange, cations and anions trade places in a correlated process somewhat similar to simultaneous, multiple ring diffusion. Breaking of bonds internal to the oxide can be a field-assisted process giving rise to potentially mobile ions. Place exchange with neighboring ions could then occur in a chain reaction such that cations would move toward the oxide–gas interface and anions toward the oxide–metal interface. Fromhold [25] holds that simultaneous place exchange of several pairs of ions is statistically unlikely, so that a non-simultaneous process is favored. This process is illustrated in Figure 4.4.

It is actually surprising that both cations and anions simultaneously move in an oxide since energies of formation of cation and anion vacancies

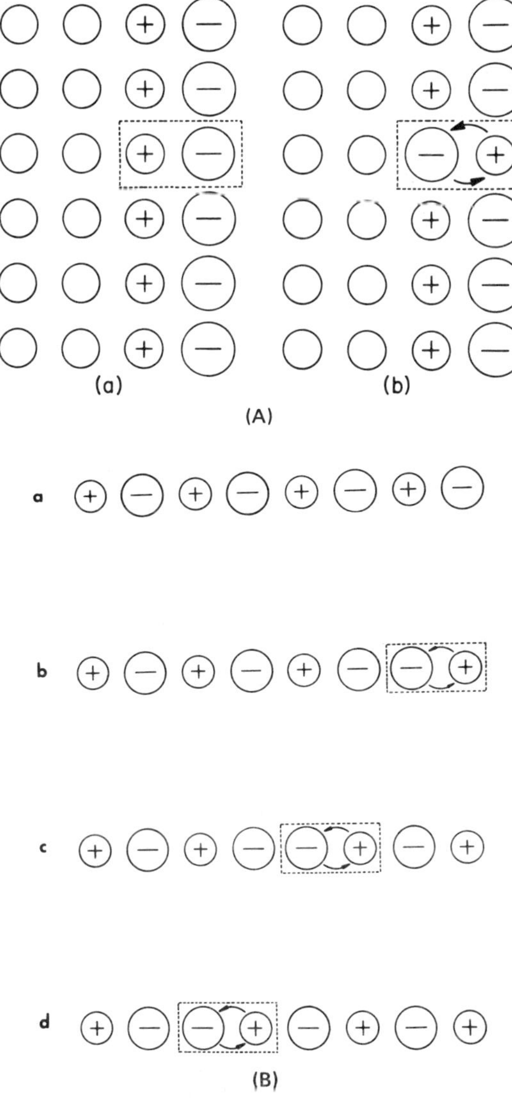

Figure 4.4. Mechanism of place exchange: (A) permutation of cation and anion positions in a dipolar surface layer (a) before and (b) after; (B) nonsimultaneous place-exchange events proceeding from (a) through (d). After Fromhold [25]. Reprinted by permission of the publisher, The Electrochemical Society, Inc.

in crystals vary [50]. It would seem that the energy barrier U for movement of an anion through a vitreous oxide would be much greater than that for a cation as in many crystalline oxides, even though the energy barrier for entry W is less ($U_A \gg U_C$, $W_A < W_C$). See Figure 1.1. However, in a noncrystalline body, U_A may be only slightly greater than U_C, so given $W_A < W_C$, the sum $W_i + U_i$ can be almost the same for both species. It is this sum which determines the rate of oxidation.

According to our first model, if both cations and anions were mobile in an oxide, then a high probability of collision would lead to oxide formation within the channel, closing it up. A brief consideration of the probability of such an occurrence indicates that it is not very likely to happen [51]. This conclusion agrees with experiment [52].

The very postulate of channels recalls the oxidation model of Evans [53]. In this model, ions move through pores or grain boundaries in the oxide. Gradual filling of the pores with oxide results in a slowing of ion movement due to an increase in the activation energy. Eley and Wilkinson, following Lanyon and Trapnell [48], likewise assumed an increase in activation energy for ion movement as film growth proceeded by a place exchange mechanism. Direct logarithmic kinetics resulted from integrating

$$\frac{dx}{dt} = cN \exp\left(-\frac{W + \lambda' x}{kT}\right) \tag{9}$$

where x is the oxide thickness, c, λ' are constants, and N is the concentration of potentially mobile ions.

This discussion lends a note of caution to the interpretation of oxidation models. If indeed, ion movement in network-forming oxides occurs because of the open structure or via other paths of easy-ion motion, then it is possible that the morphology of the paths can change with time. Constriction of channels would lead to an increase in the barrier to ion movement W as pointed out by Fehlner and Mott [47]. This would result in direct logarithmic kinetics. If channels were to close up completely, then N in the preexponential of Equation 9 would decrease. Examples of W increasing with oxide thickness have been recorded in the literature [54,55]. If the effect is large enough, it could overshadow the inverse logarithmic kinetics of the Cabrera–Mott expression.

Conversely, channels through an oxide may increase in both number and size if oxide rearrangement occurs and crystallites form. The defects or grain boundaries so formed provide paths of easy-ion movement [56] since the diffusing species are no longer forced to squeeze between atoms constrained by a vitreous network or single crystal structure. Instead,

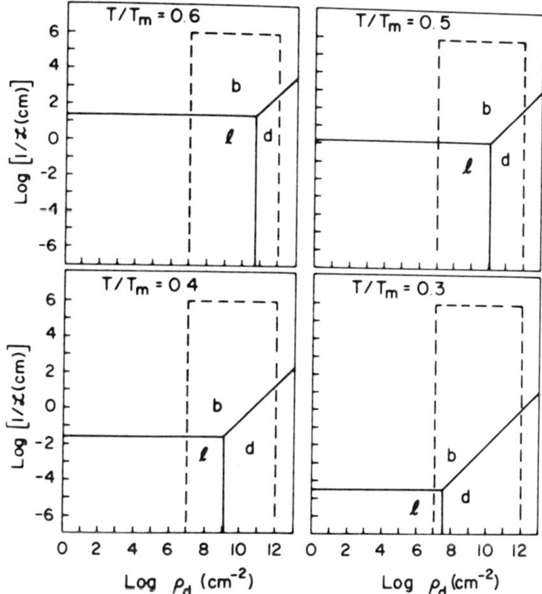

Figure 4.5. Regimes of reciprocal grain size $1/\mathcal{L}$ and dislocation density ρ_d over which lattice diffusion (l), grain boundary short-circuit diffusion (b), or dislocation short-circuit diffusion (d) is dominant during steady-state diffusion through a thin film specimen of an f.c.c. metal. Each map corresponds to diffusion at a temperature T which is a fixed fraction of the absolute melting temperature T_m. After Balluffi and Blakely [26]. Published by permission of the copyright holder, Elsevier Sequoia.

diffusion occurs along grain boundary surfaces where the activation energy for movement is less. For example, oxidation of polycrystalline silicon [57] is enhanced in the region of grain boundaries. A linear dependence of ln $\overline{C}$ on x is often found for grain boundary diffusion, where $\overline{C}$ is the average diffusant concentration as a function of oxide depth x. This contrasts with the Fick's law relationship, ln $\overline{C}$ versus x^2, found for bulk diffusion. The presence of impurities at grain boundaries can greatly influence the experimental observations.

Balluffi and Blakely [26] have reviewed diffusion in thin films, showing under what conditions dislocation and grain boundaries can dominate for crystalline materials. The simplest case to illustrate is that of an f.c.c. metal as shown in Figure 4.5. Four diffusion maps based on grain size $\mathcal{L}$ and dislocation density ρ_d are outlined. Lattice diffusion l, grain boundary short-circuit diffusion b, or dislocation short-circuit diffusion d are defined under steady state conditions. The variable T/T_m for each map is the dif-

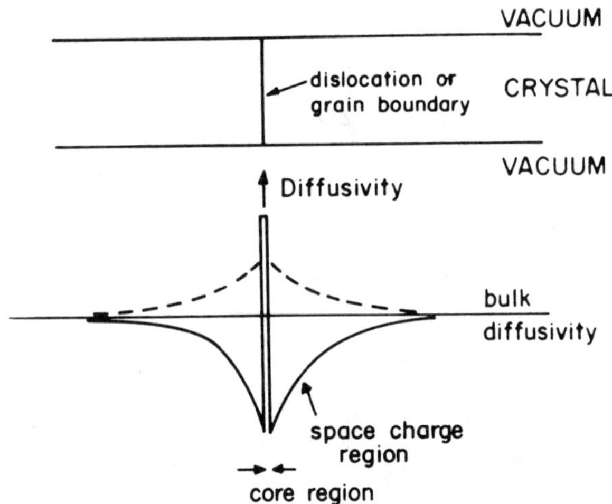

Figure 4.6. Possible form of the diffusivity as a function of position near a grain boundary or dislocation in an ionic solid. Two regions are distinguished: the core and the space charge regions. In the latter, the diffusivity parallel to the boundary may be enhanced (broken curve) or suppressed (solid curve). After Balluffi and Blakely [26]. Published by permission of the copyright holder, Elsevier Sequoia.

fusion temperature T as a fixed fraction of the absolute melting point T_m. The diffusion fluxes were calculated from diffusion parameters proposed by Gjostein [58]. It is apparent from Figure 4.5 that short-circuit diffusion should dominate in the presence of grain boundaries.

Diffusion in oxide films which have crystallized to form grains, for example, in modifying oxides or impure vitreous oxides, is similar in principle to diffusion in metals, but must be modified by ionic influences represented by the Debye length [59]. This length is the characteristic decay distance for the space charge and depends on temperature, dielectric constant, and impurity level. It can be of the order of 0.1–1 μm for an oxide at room temperature.

The net result of such considerations is to emphasize the extended nature of defects in oxides. Dislocations or grain boundaries containing localized ions can influence the motion of other ions in the surrounding oxide. Movement of these ions will be either enhanced or suppressed, depending on their mode of transport (interstitial or vacancy in crystals) and the magnitude of the space charge. Figure 4.6 illustrates this effect. Here the grain boundary or dislocation forms a path of easy-ion movement. An anion or cation moving within this path can aid or retard the movement of surrounding ions. In this way, a correlated movement of ions can occur. This effect is, however, expected to be small if the space charge density is small as in low-temperature oxidation.

The role of impurities in diffusion through oxides should be brought into the discussion at this point. The role of dopant ions in changing the rate of oxidation of a metal at high temperatures where diffusion through crystalline oxide dominates is well documented [60]. If the valence of the dopant ion varies from that of the host ion, then the number of vacancies in an ionically bonded crystalline oxide will change to maintain electrical neutrality. Thus, the number of ions moving by a substitutional mechanism is increased or decreased.

There are indications that the same effect of dopant ions on oxidation is true at low temperatures when the oxide formed is a polycrystalline modifier [61]. However, the model does not appear to apply to vitreous oxide networks. Instead, it is proposed that impurity ions, including water, can act as a modifier [62]. Water appears to diffuse in oxide glasses as molecules [63], reacting to break up the network structure, especially at vulnerable points such as channels. Hydroxyl groups are formed which weaken the network, providing points for ion exchange or for further field-assisted bond breaking. Thus, movement of network ions by diffusion, drift, or place exchange is aided. On the other hand, impurities such as water can strengthen a weak oxide if additional bonding in the form of hydrogen bonding is introduced. Thus, water can aid the formation of vitreous chromium oxyhydroxide [64] and other metal oxyhydroxides.

As pointed out above, formation of grain boundaries can dramatically alter oxidation rates. Impurities are known to be an aid to recrystallization and hence speed the formation of grain boundaries. In fact, oxidation of silicon is considered to be an impurity-controlled process [65,66], at least in the parabolic regime. Certain other impurities such as carbon do not enter the oxide but remain at the metal surface where they interfere with the early stages of metal oxide growth, that is, the logarithmic or linear regimes [67,68], and may change the oxide morphology. See Chapter 5.

Studies of anodic oxides provide further insight into the movement of network ions during low-temperature oxidation. As pointed out above, ion movement during anodization is a field-assisted process, as is low-temperature oxidation, so that the analogy between the two can be closely drawn.

Table 4.1 taken mainly from Pringle [52] includes the results of marker studies on anodic oxides. An impurity ion implanted in the oxide is detected by radioactive isotope sectioning, Rutherford backscatter analysis, or nuclear microanalysis. The location of the implanted ion reveals the ratio of cations to anions which are mobile in the oxide, dependent on

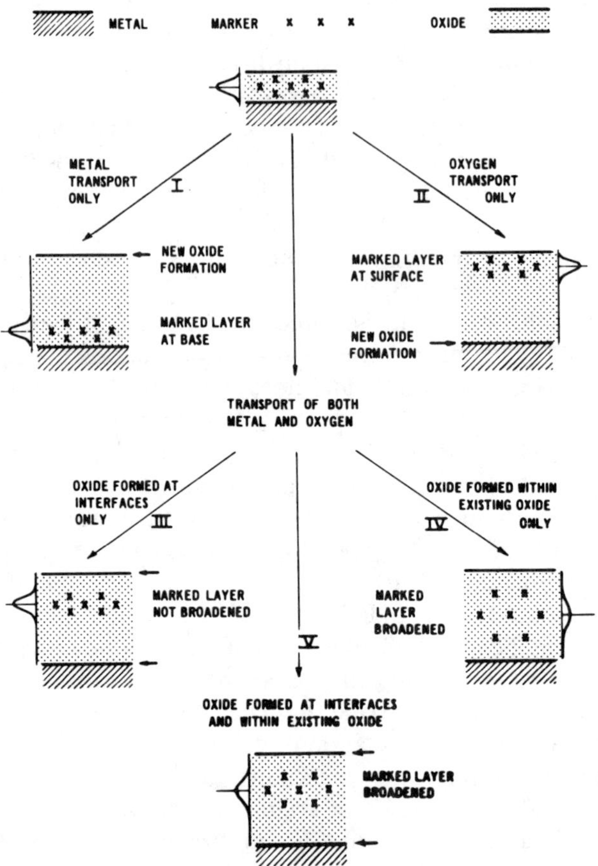

Figure 4.7. Effect of metal and oxygen migration on the position of inert, immobile markers in an oxide film. The resulting marker profiles are shown at the side of each diagram. After Pringle [69]. Reprinted by permission of the publisher, The Electrochemical Society, Inc.

the metal, electrolyte, and current density. Figure 4.7 outlines the various possibilities which can be differentiated using markers [69].

Noble gas markers are best [70] since they carry no charge when in the oxide. Charged markers, on the other hand, tend to move with the growing oxide. Motion of such markers depends partly on the nature of the charge, for example, alkali ions in aluminum oxide move toward the oxide–gas interface with mobility K > Pb > Cs while halogens move toward the metal–oxide interface with mobility Cl > Br > I. Movement is also modified by whether the tracer, for example, As, Sb, Bi, Se, Te, is implanted in the aluminum or in the oxide [71]. A migration number can be defined as the ratio of impurity marker movement to thickness of oxide grown. During the anodization of Si, phosphorus was found to move inward [72]. Bromine migrates inward during tantalum anodization, while Rb moves outward [73].

Metal dissolution during anodization has been a subject of concern since it may affect measurements of oxide growth rate. For the anodized metals contained in Table 4.1, only V and Al show extensive dissolution in water [74].

Both field and electrolyte can affect transport numbers determined using inert gas markers. For instance, the cationic transport number in Al_2O_3 varies from 0.35 at 0.1 mA cm^{-2} to 0.72 at 10 mA cm^{-2} in 3% aqueous ammonium citrate. It remains at ~0.6 for the same current density range in tetraborate–glycol [75].

The most extensive study to date on any single metal has been carried out by Pringle [69]. He has carefully analyzed the anodization of tantalum and for the variables examined, has concluded that:

1. Both anions and cations move with transport numbers of 0.757 and 0.243, respectively, at 25°C and a current density of 1 mA cm^{-2} in 0.1 *M* H_2SO_4.
2. New oxide growth occurs only at the oxide interfaces.
3. Anions and cations move under conditions of forced diffusion by successive single jumps so that their order in a queue is maintained.
4. A "Brownian" motion of the anions and cations occurs simultaneously with movement in the queue.

The above conclusions were obtained through the use of two classical techniques associated with the study of oxide growth; markers and tracers. Conclusions 1, 2, and 4 came from a study of the location of radioactive noble gas markers in anodic tantalum oxide [75]. Possibility III in Figure 4.7 was chosen as the mechanism of anodic oxide growth on tantalum.

Conclusion 3 for anions [76] and cations [77] resulted from a study of the fate of ^{18}O and ^{182}Ta tracers during anodic oxidation. In this case, the actual ions which move during oxidation served as a probe of events occurring in the process. Figure 4.8 outlines the possibilities which can be differentiated by the tracer technique. There are, indeed, other possibilities, especially combinations of ions, but these have been ruled out for tantalum oxide by results from the marker studies.

Verkerk et al. [78] used a radioactive tantalum marker to confirm that the order of tantalum atoms is largely preserved during anodization. It has been shown using nuclear microanalysis [79] that order is also con-

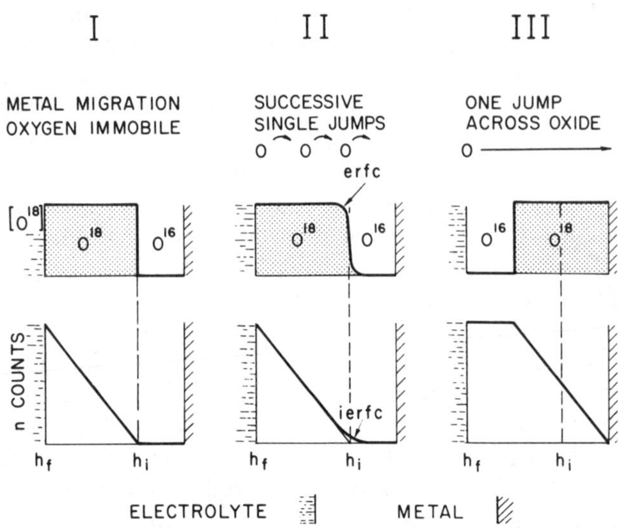

Figure 4.8. Three basic ^{18}O tracer concentration profiles in anodic tantalum oxide when the oxide is formed initially to a thickness h_i in $H_2{}^{16}O$ and then to h_f in $H_2{}^{18}O$. The lower curves show the corresponding neutron count profiles when the films are analyzed by the ^{18}O (p, n) ^{18}F method. After Pringle [76]. Reprinted by permission of the publisher, The Electrochemical Society, Inc.

served during the anodization of aluminum, provided compact oxide forms. When porous Al_2O_3 is formed, the oxygen order inverts [80] due to oxygen permeation down paths of easy-ion movement.

Leach and Neufeld [81] have remarked upon the surprising plasticity of oxide films which are undergoing anodization. The implication of both the "Brownian" motion (Pringle's conclusion 4 above) and oxide flexibility is that there are a great many broken bonds in the oxide network during the time that a high field is being applied, especially when this field is a significant fraction of the oxide bond strength. As a result, ion movement is made much easier. Broken bonds due to impurities or modifying oxides contribute to easier ion movement as pointed out above.

It is important in the theory of oxidation to explain, not only why both anions and cations can move, but also how the ratio of the two changes with ion flux. It has been found that, in the case of aluminum (Table 4.1) anodized in aqueous ammonium citrate, a larger fraction of the ionic current is carried by cations as the field is increased. An explanation for this

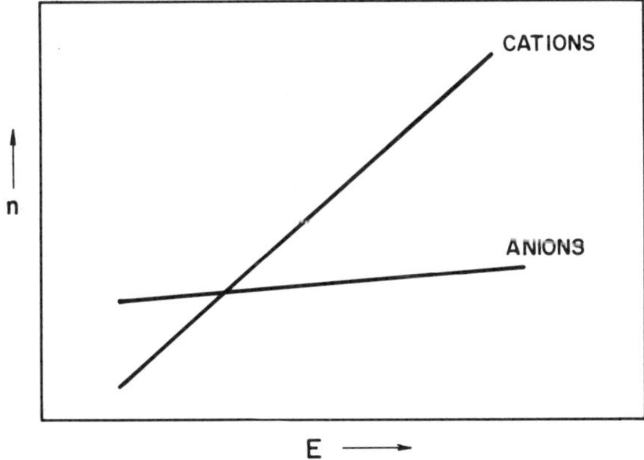

Figure 4.9. Proposed explanation for the change in cation–anion transport numbers as a function of applied field for anodic oxides.

effect has been offered [82]. If one considers separate equations describing anion and cation movement such that

$$n_A = n_A^0 \exp\left(-\frac{W_A - qaE}{kT}\right) \tag{10}$$

$$n_C = n_C^0 \exp\left(-\frac{W_C - qaE}{kT}\right) \tag{11}$$

where n^0 is the concentration of ions available to move from an interface, $W_{A,C}$ is the activation energy for ion motion, and q is the ionic charge, then it is possible for a crossover in n_A, n_C to occur as a function of field, provided that the larger n^0 is associated with the smaller W. See Figure 4.9.

A similar effect is seen in the temperature dependence of bipolar conductivity [83]. When only one ion is moving, a plot of log conductivity versus inverse temperature yields a straight line. In the bipolar case, two straight lines are found.

We previously discussed the simultaneous injection of cations and anions at oxide interfaces as being the explanation for movement of both species through an oxide. A second explanation for simultaneous anion and cation movement has been advanced by Amsel and Samuel [79] who suggested a mechanism which involves the breaking of oxide bonds within

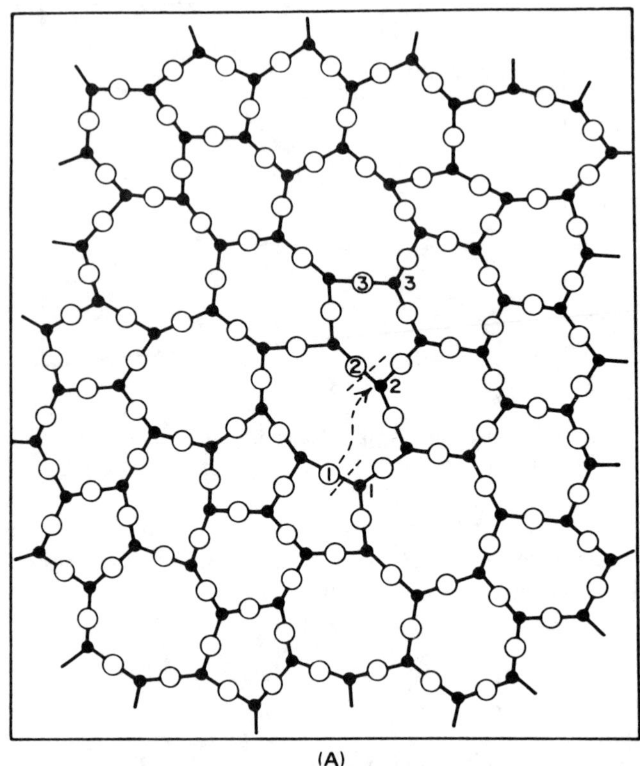

Figure 4.10. Random-network model for vitreous B_2O_3 showing: (A, B) defect formation and (B, C) defect migration through the network. Solid circles are boron atoms while open ones are oxygens. In producing these figures, the atoms at the periphery were maintained fixed and the remaining atoms moved so as to maintain approximately constant bond lengths and angles. After Dignam. Reprinted from Ref. 41, pp. 196–197, by courtesy of Marcel Dekker, Inc.

the oxide. This idea is similar in some respects to the place exchange model for ion movement as discussed by Fromhold, and to the Frenkel defect model put forward to explain transients during anodization.

A variant of these models has been proposed by Dignam [41]. His approach depends on the characteristics of vitreous oxides. Under the influence of a field or temperature, defects are postulated to form in the oxide. For odd-valent metal oxides, a pair of defects is envisaged, separating through a succession of partner exchanges. Even-valent metal oxides are thought to produce only displaced oxygen ions.

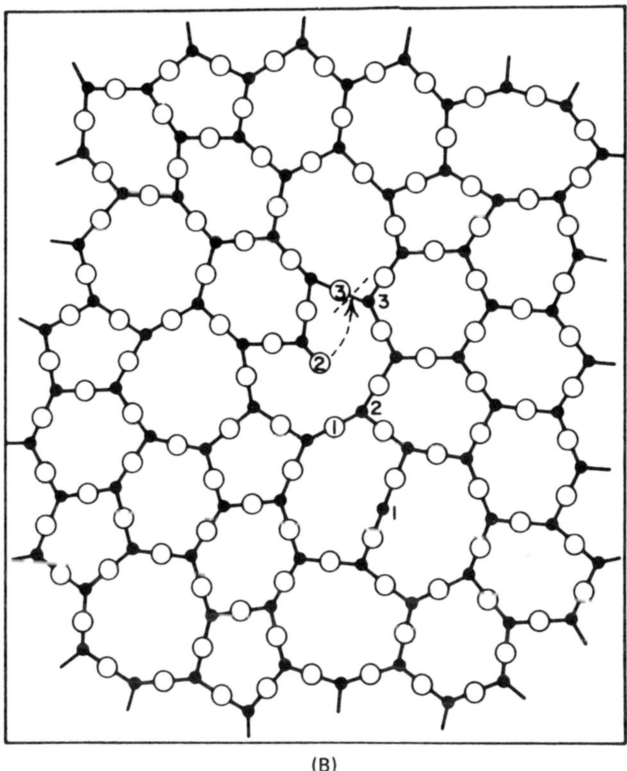

(B)

Figure 4.10. (*Continued*)

Movement of the defects can occur as illustrated in Figure 4.10. A cooperative movement of the network in the vicinity of a defect pair allows the dangling bonds to move through the network. The displaced network atoms do not necessarily relax to their original positions after passage of the defect since the very existence of the vitreous state in part depends upon multiple configurational states of equivalent energy [45]. The movement of the defects through the network accounts for several experimental observations: the movement of both anions and cations, the maintenance of order, and the formation of new oxide only at interfaces.

Dignam has calculated the enthalpy of formation ΔH_d^0 of network-defect pairs. Approximations are made and the enthalpy of hydration used to estimate values. Results are given in Table 4.2 based on high-frequency AC dielectric constants. The equilibrium concentration of defects would

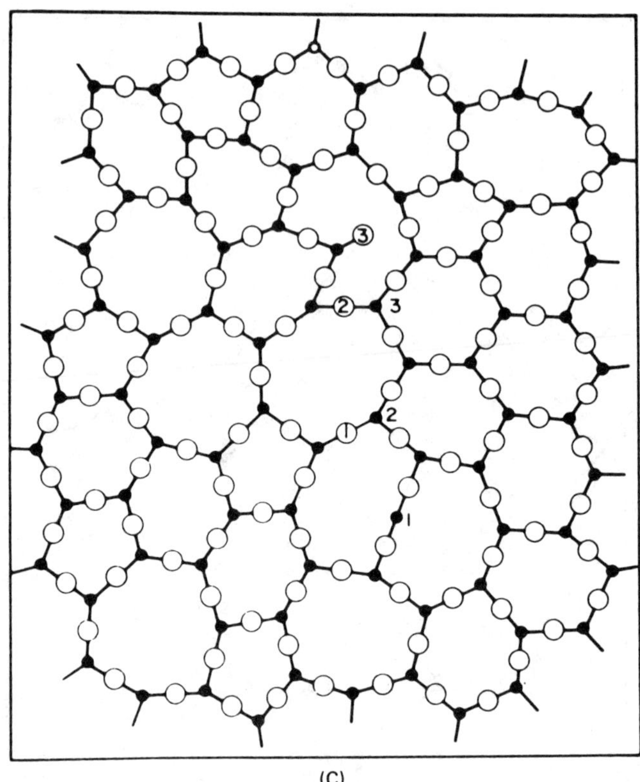

(C)

Figure 4.10. (*Continued*)

be proportional to $\exp[-\Delta H_d^0/2kT]$ and the low field activation energy for ionic conduction would be $[\Delta H_d^0/2 + U]$ where U is the energy barrier to ion movement and is approximately equal to $\Delta H_d^0/2$. Thus, the activation energy is approximated by ΔH_d^0. Results for B_2O_3 are in fair agreement with this model.

Calculations indicate that Frenkel and Schottky defects in odd-valent crystalline oxides have a much higher ΔH_d^0 than network defects in vitreous oxides. Thus, ion movement by a network-defect mechanism should be favored over an interstitial or vacancy mechanism. The structure of the oxide is related to the mechanism of ionic transport and in turn, the network-defect mechanism may encourage the formation of vitreous oxide under favorable circumstances such as anodization.

Mott [84] has proposed that comparable transport numbers for anions

TABLE 4.2. ESTIMATED ENTHALPY OF FORMATION FOR NETWORK-DEFECT PAIRS [41]

Vitreous oxide	ΔH_d° (eV)
B_2O_3	2.28
Al_2O_3	1.21
Bi_2O_3	0.61
Ta_2O_5	3.43
Nb_2O_5	2.19

and cations in vitreous oxides would occur if ion movement through vitreous oxides could be looked at as a ''melting'' phenomenon. A correlated rearrangement of several atoms may occur under the influence of thermal fluctuations such that a net movement of ions occurs with the order of moving species preserved. The activation energy for the process should approximately equal nL' where n is the number of atoms involved and L' is the latent heat for the glass transition. Values of L' are in the range of 0.1–1 eV so that for an activation energy of 1 eV, $n \cong 1$ to 10. Provided the activated state results in a charge displacement of qa, then an applied field would reduce nL' by qaE. Ion movement according to this mechanism would result in mixing of inert marker atoms such that the marker broadening equals $a_m n^{1/2}$ where a_m is the number of ''melting'' events per atom which must occur to produce n layers. Pringle [69] has shown that such a relationship is indeed found for the anodization of tantalum, that is, the broadening of inert gas markers varies as the square root of the thickness of oxide formed, although he interprets it differently.

These latter models lead to similar explanations for the following experimental observations:

1. Potentially mobile anions and cations are formed in the oxide so there is no need for simultaneous ion incorporation at the two interfaces.
2. New oxide is formed at the oxide interfaces when the mobile ions reach their respective interfaces and react.
3. The distribution of a rare gas marker is not greatly broadened or skewed because the rare gas atoms can fall back into holes left by mobile ions moving toward the interfaces.
4. Ion order is maintained.

5. Broken bonds and a cooperative disturbance involving exchange account for the observed "Brownian" motion of Pringle.
6. The increase of oxide growth rate and porosity due to the absorption of ultraviolet light by the oxide [85] can be ascribed to breaking of oxide bonds located within the oxide, although other explanations are possible.

It is not at all apparent that any models discussed above totally explain the movement of network-forming ions in a vitreous oxide. One major difference between current models is the source of the mobile species; incorporation at an interface or generation within the oxide. It is certainly possible that specific oxides fall under one or the other model, depending on bond energies and oxide structure. Assignment of a model to a specific oxide must await further experimental work.

We have then a partial picture of ion movement in vitreous oxides grown under field at low temperatures. The picture is in many ways incomplete, but may be summarized as follows. Mobile anions/cations are generated in the oxide either at one or both interfaces or by breaking internal bonds. The ions drift across the oxide under the influence of the electric field. The process of ion movement has been depicted as either successive single jumps coupled with a "Brownian" motion or place exchange in which anions and cations trade positions in the oxide. The rate-limiting process during this sequence can be either generation of the ions, incorporation of the ions into the oxide, movement of the ions through the oxide, neutralization of the ionic charge by electronic charge, or reaction at the opposite interface. Movement through the oxide is usually taken to be the rate-limiting process.

REFERENCES

1. N. Cabrera and N. F. Mott, *Rep. Prog. Phys.* **12** (1948–49), 163.
2. R. M. Hill, *Thin Solid Films* **1** (1967), 39; A. K. Jonscher, ibid, 213.
3. A. T. Fromhold, Jr., *Theory of Metal Oxidation, Vol. 1—Fundamentals*, North-Holland Publishing, New York, 1976; (a) p. 290; (b) p. 368.
4. B. F. Lewis, W. M. Bowser, J. L. Horn, Jr., T. Luu, and W. H. Weinberg, *J. Vac. Sci. Technol.* **11** (1974), 262.
5. N. F. Mott and E. A. Davis, *Electronic Processes in Non-crystalline Materials*, 2nd ed., Clarendon Press, Oxford, 1978.
6. J. H. Davies, *Phil. Mag.* **B41** (1980), 373.
7. R. C. Hughes, *Phys. Rev. Lett.* **30** (1973), 1333.
8. R. C. Hughes, *Appl. Phys. Lett.* **26** (1975), 436; *Phys. Rev.* **B15** (1977), 2012.
9. B. T. Kolomiets, *Phys. Stat. Solidi* **7** (1964), 359, 713.
10. W. E. Spear et al., *Solid State Commun.* **17** (1975), 1193; *Appl. Phys. Lett.* **28** (1976), 105.
11. P. W. Anderson, *J. de Physique* **C-4** (1976), 339.

12. D. L. Griscom, in *The Physics of SiO_2*, S. T. Pantelides, Ed., Pergamon Press, New York, 1978, p. 232.
13. N. F. Mott, *Phil. Mag.* **19** (1969), 835.
14. P. W. Anderson, *Phys. Rev. Lett.* **34** (1975), 953.
15. R. A. Street and N. F. Mott, *ibid.* **35** (1975), 1293.
16. N. F. Mott, E. A. Davis, and R. A. Street, *Phil. Mag.* **32** (1975), 961.
17. M. Kastner, D. Adler, and H. Fritzsche, *Phys. Rev. Lett.* **37** (1976), 1504.
18. N. F. Mott, in *The Physics of SiO_2*, S. T. Pantelides, Ed., Pergamon Press, New York, 1978, p. 1.
19. N. F. Mott, *Adv. Phys.* **26** (1977), 363.
20. R. A. Street, T. M. Searle, and I. G. Austin, *Phil. Mag.* **29** (1974), 1157.
21. N. Mott, M. Pepper, S. Pollitt, R. H. Wallis, and C. J. Adkins, *Proc. Roy. Soc.* (*London*) **A345** (1975), 169.
22. M. Pepper, *ibid.* **A353** (1977), 225.
23. Per Kofstad, *Nonstoichiometry, Diffusion and Electrical Conductivity in Binary Metal Oxides*, Wiley-Interscience, New York, 1972, p. 71.
24. W. Jost, *Diffusion in Solids, Liquids, and Gases*, Academic Press, New York, 1952.
25. A. T. Fromhold, Jr., *J. Electrochem. Soc.* **127** (1980), 411.
26. R. W. Balluffi and J. M. Blakely, *Thin Solid Films* **25** (1975), 363.
27. K. R. Lawless, *Rep. Prog. Phys.* **37** (1974), 231.
28. V. O. Altemose, *J. Appl. Phys.* **32** (1961), 1309.
29. F. J. Norton, *J. Am. Ceram. Soc.* **36** (1953), 90.
30. W. D. Kingery, H. K. Bowen, and D. R. Uhlmann, *Introduction to Ceramics*, 2nd ed., John Wiley & Sons, New York, 1976, p. 260.
31. A. E. Owen, in *Progress in Ceramic Science*, Vol. 3, J. E. Burke, Ed., Macmillan, New York, 1963, p. 77.
32. G. W. Morey, *The Properties of Glass*, Reinhold, New York, 1938, p. 459.
33. E. N. Randall, *Ceramic Bull.* **52** (1973), 434, Abstract No. 5-S4-73.
34. J. M. Stevels, in *Encyclopedia of Physics*, Vol. 20, S. Flügge, Ed., Springer-Verlag, Berlin, 1957, p. 350.
35. R. A. Huggins, in *Solid Electrolytes*, P. Hagenmuller and W. Van Gool, Eds., Academic Press, New York, 1978, p. 27.
36. N. F. Mott and R. W. Gurney, *Electronic Processes in Ionic Crystals*, 2nd ed., Clarendon Press, Oxford, 1948, p. 33.
37. M. S. Seltzer, in *Oxides and Oxide Films*, Vol. 4, J. W. Diggle and A. K. Vijh, Eds., Marcel Dekker, New York, 1976, p. 1.
38. M. J. Dignam, *J. Electrochem. Soc.* **126** (1979), 2188.
39. A. G. Revesz, *Phys. Stat. Solidi* **A57** (1980), 235, 657; **A58** (1980), 107.
40. C. J. Dell'Oca, D. L. Pulfrey, and L. Young, in *Physics of Thin Films*, Vol. 6, M. H. Francombe and R. W. Hoffman, Eds., Academic Press, New York, 1971, p. 1.
41. M. J. Dignam, in *Oxides and Oxide Films*, Vol. 1, J. W. Diggle, Ed., Marcel Dekker, New York, 1972, p. 91.
42. L. Young and F. G. R. Zobel, *J. Electrochem. Soc.* **113** (1966), 277.
43. L. Young and D. J. Smith, *ibid.* **126** (1979), 765.
44. M. J. Dignam, in *Comprehensive Treatise of Electrochemistry*, Vol. 4, J. O'M. Bockris, B. E. Conway, E. Yeager, and R. E. White, Eds., Plenum Press, New York, 1981, p. 247.
45. A. G. Revesz and F. P. Fehlner, *Oxid. Metals* **15** (1981), 297.
46. D. L. Evans and S. V. King, *Nature* **212** (1966), 1353.
47. F. P. Fehlner and N. F. Mott, *Oxid. Metals* **2** (1970), 59.

48. D. D. Eley and P. R. Wilkinson, *Proc. Roy. Soc.* (*London*) **A254** (1960), 327; M. A. H. Lanyon and B. M. W. Trapnell, *ibid.* **A227** (1955), 387.
49. M. Cohen, *J. Electrochem. Soc.* **121** (1974), 191C.
50. Z. A. Munir and J. P. Hirth, in *Surfaces and Interfaces in Ceramic and Ceramic-Metal Systems*, J. Pask and A. Evans, Eds., Plenum Press, New York, 1981, p. 23.
51. A. T. Fromhold, Jr., in *Passivity of Metals*, R. P. Frankenthal and J. Kruger, Eds., The Electrochemical Society Inc., Princeton, NJ, 1978, p. 59.
52. J. P. S. Pringle, *Electrochim. Acta* **25** (1980), 1403, 1423.
53. U. R. Evans, *Metallic Corrosion Passivity and Protection*, 2nd ed., Edward Arnold, London, 1946.
54. H. Heyne and F. C. Tompkins, in *Surface Phenomena of Metals*, S.C.I. Monograph No. 28, Society of Chemical Industry, London 1968, p. 339.
55. W. A. Crossland and H. T. Roettgers, *Phys. Failure Electron.* **5** (1967), 158.
56. J. W. Cathcart, G. F. Petersen, and C. J. Sparks, in *Surfaces and Interfaces I*, J. J. Burke et al., Eds., Syracuse University Press, Syracuse, NY, 1967, p. 333.
57. L. L. Kazmerski, O. Jamjoum, P. J. Ireland, and R. L. Whitney, *J. Vac. Sci. Tech.* **18** (1981), 960.
58. N. A. Gjostein, *Diffusion*, American Society for Metals, Metals Park, 1973, p. 241.
59. S. Glasstone, *Physical Chemistry*, 2nd ed., D. Van Nostrand Co., New York, 1946, p. 958.
60. O. Kubaschewski and B. E. Hopkins, *Oxidation of Metals and Alloys*, 2nd ed., Butterworths, London, 1962.
61. P. K. Krishnamoorthy and S. C. Sircar, *J. Electrochem. Soc.* **116** (1969), 734; *Acta Met.* **16** (1968), 1461; **17** (1969), 1009; *Corrosion* **24** (1968), 407; *Scripta Met.* **2** (1968), 255; *Oxid. Metals* **2** (1970), 349.
62. E. N. Boulos and N. J. Kreidl, *J. Can. Ceram. Soc.* **41** (1972), 83.
63. R. H. Doremus, *Glass Science*, John Wiley & Sons, New York, 1973, p. 134ff.
64. G. Okamoto and T. Shibata, in *Passivity of Metals*, R. P. Frankenthal and J. Kruger, Eds., The Electrochemical Society, Princeton, NJ, 1978, p. 646.
65. A. G. Revesz and R. J. Evans, *IEEE Trans.* **ED-14** (1967), 789; *J. Phys. Chem. Solids* **30** (1969), 551.
66. F. P. Fehlner, *J. Electrochem. Soc.* **119** (1972), 1723.
67. P. B. Needham, Jr., H. W. Leavenworth, Jr., and T. J. Driscoll, *ibid.* **120** (1973), 778.
68. P. B. Sewell, D. F. Mitchell, and M. Cohen, *Surface Sci.* **33** (1972), 535.
69. J. P. S. Pringle, *J. Electrochem. Soc.* **120** (1973), 398, 1391.
70. F. Brown and W. D. Mackintosh, *ibid.* **120** (1973), 1096.
71. W. D. Mackintosh, F. Brown, and H. H. Plattner, *ibid.* **124** (1977), 1168.
72. W. D. Mackintosh and H. H. Plattner, *ibid.* **124** (1977), 396.
73. J. L. Whitton, *ibid.* **115** (1968), 58.
74. J. Siejka, J. P. Nadai, and G. Amsel, *ibid.* **118** (1971), 727.
75. J. A. Davies, B. Domeij, J. P. S. Pringle, and F. Brown, *ibid.* **112** (1965), 675.
76. J. P. S. Pringle, *ibid.* **120** (1973), 1391.
77. J. P. S. Pringle, *Electrochem. Soc. Ext. Abstr.* **78-1**, Abstr. No. 195, Seattle, WA, May 1978.
78. B. Verkerk, P. Winkel, and D. G. de Groot, *Philips Res. Rep.* **13** (1958), 506.
79. G. Amsel and D. Samuel, *J. Phys. Chem. Sol.* **23** (1962), 1707.
80. C. Cherki and J. Siejka, *J. Electrochem. Soc.* **120** (1973), 784.
81. J. S. L. Leach and P. Neufeld, *Corrosion Sci.* **9** (1969), 225.
82. P. A. Brook, V. R. Howes, J. S. L. Leach, and A. Y. Nehru, Paper 94, Abstracts 138th Nat'l Meeting, The Electrochemical Society, October 1970, p. 242.

83. A. L. G. Rees, *Chemistry of the Defect Solid State*, Methuen and Co., London, 1954, p. 85.
84. N. F. Mott, private communication.
85. L. Young, *Anodic Oxide Films*, Academic Press, New York, 1961, pp. 135–139.
86. J. Bardeen, W. H. Brattain, and W. Shockley, *J. Chem. Phys.* **14** (1946), 714.
87. P. K. Krishnamoorthy and S. C. Sircar, *Acta Met.* **16** (1968), 1461.
88. M. T. Shehata, C. J. Good-Zamin, and R. Kelley, Proc. Symp. Thin Film Phenomena, The Electrochemical Society, Princeton, NJ, 1978, p. 78.
89. W. D. Mackintosh and H. H. Plattner, *J. Electrochem. Soc.* **123** (1976), 523.

5

INITIAL OXIDE FORMATION

5.1. PROCESSES OCCURRING DURING OXIDATION OF A CLEAN METAL SURFACE

A totally clean metal surface is seldom if ever found in nature and must be prepared if needed for experimental work. It is instructive to consider the processes which can take place when such a surface is exposed to oxygen. They may be summarized as follows [1,2] assuming a quasi-equilibrium of electrons so that ionic processes are rate limiting:

1. Oxygen is adsorbed on the clean metal surface to form a chemisorbed monolayer. The sticking coefficient for oxygen in this process is often high. The heat of adsorption is also high, close to the value for the heat of reaction.
2. Additional layers of oxide build up, usually via island growth, possibly through a process of place exchange in which it is postulated that the ion image force reduces the activation energy for ion movement. Processes 1 and 2 give rise to linear kinetics directly dependent upon oxygen pressure.
3. A stable oxide layer forms. Electrons transported through the oxide layer are captured by electron traps at the oxide–gas interface, building up voltage V across the oxide. Hole transport is also possible, depending on the metal and oxide band structures. The traps are due to adsorbed oxygen on the oxide surface. The resulting charge sets up an electric field E across the oxide such that the activation energy for ion motion is reduced to $W - qaE$, where W is the energy barrier to ion movement, q is the ionic charge, and $2a$ is the distance between stable ion positions.
4. Either cations, anions, or both can move as influenced by the crystalline or vitreous structure of the oxide. At this point in the oxidation process, it is proposed that an essentially constant voltage exists across the oxide so that an increase in oxide thickness leads to a decreasing field.

Processes 1 through 4 are observed at low temperatures where they proceed slowly enough to follow using measurement techniques having a few seconds response time. At higher temperatures, thermally activated oxidation driven by a chemical gradient overshadows steps 3 and 4. Often, linear kinetics based on interface reaction or parabolic kinetics based on diffusion processes in the oxide may then be observed. The temperature at which the transition from low- to high-temperature oxidation occurs can be a function of the metal and its purity as well as oxygen pressure.

Those processes which occur during initial oxide growth are discussed more fully below.

5.2. ADSORPTION ON A CLEAN METAL

A truly clean metal surface suitable for studying the adsorption of gases is difficult to achieve. Metals other than gold [3] exposed to the atmosphere are covered with a tightly bonded layer of oxygen which can include impurities segregated to the metal surface. Cleaning of such a surface is time consuming and requires careful technique combined with sensitive analytical methods [4,5].

Generation of a clean metal surface in ultrahigh vacuum is an effective technique, provided the resulting geometry of the metal is acceptable. Polycrystalline films deposited on a substrate by evaporation or sputtering are ideal pristine surfaces which can be generated *in situ*. They are technologically useful in electronics and as such their study is doubly valuable. Films are especially useful in determining heats of chemisorption.

Crushing or cleaving large crystals *in vacuo* is another useful method. The crystal face exposed is determined by the cleavage plane of the crystal, thus limiting the faces available for study.

Metal surfaces which are already contaminated require special treatment. An oxide which is volatile at high temperatures can be removed from a metal such as tungsten or molybdenum by flash heating in vacuum [6]. However, all the oxygen may not desorb. Some may dissolve in the metal [7]. Chemical reduction of an oxide with molecular or atomic hydrogen is effective. The adsorbed hydrogen must subsequently be removed by heating in vacuum. Sulfur and carbon are removed by an oxidative treatment. A balance must be struck between oxide formation and sulfur or carbon removal. Physical bombardment of a surface with inert gas ions or neutrals also results in cleaning, in this case by sputtering. The ions or neutrals transfer kinetic energy to surface atoms which in turn leave the surface. Inert gases do not adsorb on the metal although they can be driven into the surface. This implanted gas is removed by annealing in vacuum. Heating also restores the crystallinity of the metal surface which was disrupted by the bombarding gas. A discussion of surfaces and adsorption processes is to be found in Roberts and McKee [8].

Surface atoms have fewer neighbors than atoms in the bulk. The resulting asymmetry in forces causes surface atoms to behave differently. For instance, LEED studies [9] indicate that nickel atoms on (110) and (111) surfaces are displaced from the next atomic layer about 0.1 Å more than the normal lattice spacing. The amplitude of surface atom vibrations is also reported to be greater than that of bulk atoms [10,11]. Silicon and germanium [12,13,14], as well as some clean metal surfaces, for example, Au, Pt, Ir, Mo, and W [15], undergo a rearrangement which yields structures differing from those in the bulk.

The formation of an adsorbed surface layer of gas on a metal is the beginning of the reaction between the two [16]. Approximately 10^{15} sites per square centimeter exist at a surface so that a monolayer of gas bombards the surface every 2 sec at 10^{-6} Torr and room temperature. Thus, ultrahigh vacua of 10^{-9} to 10^{-10} Torr are required for the exacting study of oxygen adsorption using LEED and other modern techniques.

Two types of adsorption [17,18] have been distinguished during the course of initial reaction: physical and chemical [19,20,21]. In physisorption, gases are weakly bound to the surface by van der Waals forces. On the other hand, a chemical bond forms between the gas and metal in chemisorption. Sometimes, a difference in the heat of adsorption can be used to distinguish between the two.

Chemisorption may proceed slowly when it must overcome an activation energy. As such, it requires higher temperatures than physisorption. The chemisorption of gases on clean metals at room temperature is very rapid, indicating that the magnitude of the activation energy is small. A supply of atomic oxygen helps overcome this barrier, leading to strong reaction in the case of tungsten [22].

Physisorption tends to occur below the boiling point of the adsorbed gas, proceeding with no activation energy at a rate proportional to the flux of gas molecules hitting the surface. It is nonspecific, as contrasted with chemisorption where crystallographic orientation and surface-active sites affect the rate. Chemisorption of a gas on a metal can proceed to the extent of one or more monolayers. These in turn can be covered by physisorbed gas.

Chemisorption can result in the formation of a strong dipole at the metal surface [23]. A decrease in surface potential (increase in work function) is experienced due to the shift of electrons away from the metal. See Chapter 6 where experimentally measured surface potential changes are collected in Table 6.1. The change in surface potential is reflected by a change in work function which is equal in magnitude but opposite in sign to the change in surface potential.

Physisorption usually has a smaller effect on surface potential than chemisorption, tending to increase it by a small amount. Exceptions are known, for example, Xe on W. As a result, both positive and negative changes have been found.

5.2.1. Nature of Adsorbed Species

The nature of the oxygen species adsorbed on a metal or oxide has long been a subject of inquiry. The use of infrared (IR), electron spin resonance (ESR), photoelectron spectroscopy (PES) based on X-rays (XPS) or ultraviolet light (UPS), low-energy electron diffraction (LEED), Auger spectroscopy, and other surface analytical techniques [24,25,26,27,28,29] have provided a means for identifying the adsorbed species. Historically, studies on nitrogen and carbon monoxide adsorption served as an aid in defining the status of oxygen. Hence, adsorption of these gases is reviewed prior to treating oxygen.

One of the first systems to be extensively analyzed was based on nitrogen. Flash desorption coupled with other techniques [30,31] provided the means for qualitatively identifying different binding states. The γ, α, and β states were found to have binding energies of 9, 20, and 81 kcal $mole^{-1}$ (0.39, 0.87, and 3.52 eV), respectively. The weakest or γ state involves molecular nitrogen which can adsorb on top of a previously chemisorbed species. The β state was thought to be a normal chemisorbed state.

Various numbers of binding states have been found for other gases: two for H_2 on W and Mo [32], two for N_2 on Mo [33], and four for CO on W [34,35]. Oxygen on tungsten is difficult to study because of surface oxide formation, but even here different binding states are suggested [36]. Further elucidation of the nature of surface sites for oxygen adsorption on (110) tungsten has been provided by Leung and Gomer [6]. A precurser state, possibly molecular, is found to exist at $T < 25°K$, which converts to a chemisorbed atomic state at 45°K. Results are correlated with the α, β, and γ states of CO adsorption on tungsten. The existence of directionally bonded adsorbates such as O_2 or H_2 on W has been confirmed by electron-stimulated desorption [37].

Oxygen adsorbed on a metal or on a stable oxide surface at room temperature converts to several subspecies. On the metal, a charged species exists as indicated by surface potential changes. On the oxide surface, both neutral O_2 and O as well as ionic species O^-, O_2^-, and O^{2-} may exist alone or in equilibrium with each other. Gland et al. [38,39] report finding molecular oxygen on Pt (111) at 120°K. The molecules dissociate into atoms at higher temperatures, illustrating the activated nature of chemisorption. On copper (100), (110), and (111) surfaces, chemisorbed molecular oxygen and weakly bonded atomic oxygen were detected between 100 and 300°K, while strongly chemisorbed atomic oxygen was identified at room temperature [40].

Some mechanisms of low-temperature oxidation beyond a monolayer ignore neutral oxygen species because they assume a field-activated process. The ions considered include O^{2-} which is postulated to move through silica at high temperatures [41]. However, O_2^- and O^- have been observed on oxide surfaces at low temperatures using electron-spin resonance [42,43,44]. ESR has also been used to study O_2 (ads), O_2^-, and O^- on TiO_2 and ZnO [45]. Formenti et al. [46] in a study of anatase particles found that the amount of oxygen adsorbed varied as $P^{1/2}$, indicating O^- as one probable species. Chemisorption and surface electronic properties of *d*-band oxides are discussed by Wolfram [28]. Eickmans et al. [29] showed, using Raman spectroscopy, that oxygen is nominally adsorbed on silver as O_2^- and O_2^{2-}.

It has been postulated [47] that an equilibrium between O_2^- and O^- can exist on the surface of titanium oxide. Both species would help to create a field during oxidation, but there is no evidence as to which species would be mobile in oxides when anion movement is significant.

Benninghoven et al. [48] have investigated surface species which can be desorbed from a metal. They used secondary ion mass spectroscopy (SIMS) and electron-induced desorption (EID). Both negative and positive ions were examined during the initial oxidation of Cr, W, V, Nb, and Ta. In all cases, the presence of O^- and O_2^- was confirmed [49].

Joyner and Roberts [50] have examined the electronic state of oxygen chemisorbed on a metal surface. The O(1*s*) binding energy was found for all metals studied to be 530.3 ± 0.1 eV referred to the Fermi level. This implies that the same or very similar species are present on all metals. Joyner and Roberts also hold that changes in surface potential can be directly related to surface coverage during the period of initial oxidation although there is some discussion of this point. See Chapter 6. The electronic states of oxygen adsorbed on oxides were found by Joyner and Roberts to vary from oxide to oxide and to differ from the electronic state of oxygen on metals.

5.2.2. Analytical Methods

The past three decades have seen the development of many methods for analyzing the surfaces of solid materials, alone or interacting with gases [15,24,25]. The principal techniques available are listed in Table 5.1 taken mainly from Somorjai's book [27]. They are applicable in general to either

TABLE 5.1. ACRONYMS USED FOR SURFACE ANALYTICAL TECHNIQUES

Technique	Acronym
Analysis of Atomic Geometry	
Low-Energy Electron Diffraction	LEED
Reflection High-Energy Electron Diffraction	RHEED
Medium-Energy Electron Diffraction	MEED
Transmission Electron Microscopy	TEM
Scanning Electron Microscopy	SEM
Scanning Transmission Electron Microscopy	STEM
Field Ion Microscopy	FIM
Low-Energy Ion Scattering	LEIS
Medium-Energy Ion Scattering	MEIS
High-Energy Ion Scattering	HEIS or RBS
Atomic Scattering and Diffraction (Surface-Sensitive Extended X-ray Absorption Fine Structure)	SEXAFS
X-ray Photoelectron Spectroscopy	XPS
Ultraviolet Photoelectron Spectroscopy	UPS
Angular Resolved Ultraviolet Photoelectron Spectroscopy	ARUPS
Ion Neutralization Spectroscopy	INS
Surface Penning Ionization	SPI
Analysis of Surface Electron Distribution	
Photoemission	
Vibrating Capacitor	
Analysis of Surface Chemical Composition	
Auger Electron Spectroscopy	AES
Scanning Auger Microscopy	SAM
Thermal Desorption Spectroscopy	TDS
Ellipsometry	
Secondary Ion Mass Spectroscopy	SIMS
Ion Scattering Spectroscopy	ISS
Rutherford Back Scattering	RBS or HEIS
Electron Induced Desorption	EID

TABLE 5.1. (*Continued*)

Analysis of Surface Vibrational Structure	
Infrared Spectroscopy	IR
Reflection Absorption Infrared Spectroscopy	RAIR
Raman Spectroscopy	
High-Resolution Electron Energy Loss Spectroscopy	HREELS
Electron Tunneling Spectroscopy	
Analysis of Surface Dynamics	
Mass Spectrometry	MS
Gas Chromatography	GC
Pressure Gauges	
Isotopic Exchange	

monolayer or multilayer oxide growth. Low energy electron diffraction, Auger electron emission spectroscopy (AES), secondary ion mass spectroscopy (SIMS), photoelectron spectroscopy as mentioned above, reflection high-energy electron diffraction (RHEED), and other techniques are reviewed in journals [51], conference proceedings [52], and textbooks [27,53,54,55]. Classical techniques such as optical spectroscopy [56] also continue to be applied to oxidation. Transmission (TEM) and scanning electron microscopy (SEM) contribute little to monolayer studies in view of their resolution limitations. However, these techniques, combined with electron diffraction, are useful in examining the morphology and state of crystallinity of both the metallic substrate and a three-dimensional oxide film. It is important to note that some methods are relatively undeveloped while others are routinely available for use.

Two techniques often used in oxidation studies are PES and AES [57]. Both analyze the atomic species present in or on a thin section of the specimen surface, of the order of 20 Å deep. They also present information on bonding and kinetics. Oxides 2–50-Å thick have been determined using these techniques [58].

Field emission microscopy (FEM) employing electrons or field ion microscopy (FIM) utilizing ions, as developed by E. W. Müller [59], have allowed direct observation of surface properties. In the field emission (electron) microscope, the specimen is a fine needle which has a hemispherical shape and a tip radius of about 1000 Å. Upon application of a high electric field, electrons are emitted from the tip and projected onto a cathodoluminescent screen, thereby creating a highly magnified image of the field distribution due to geometry and work function of the specimen surface. Adsorbed gases change the work function of the metal tip. The resulting change in pattern on the screen shows variations in different crystallographic planes. Diffusion of adsorbed atoms at cryogenic temperatures has been studied using this technique.

Studies of adsorption of oxygen on clean nickel surfaces have, for instance, shown that the emission spectra of the (110) planes are most affected. It has also been suggested from such studies that the adsorption of oxygen causes a change and rearrangement in the (110) nickel surface [30]. This conclusion is in agreement with low-energy electron diffraction studies. Oxide islands have also been observed on metals using FIM as discussed below.

The field ion microscope is used to reveal the surface structure of a metal at the atomic level. For some of the higher melting point metals, it is possible to observe single atoms on the surface. As a result, interactions and reactions between individual adsorbate and substrate atoms may be observed along with surface diffusion. Such studies have revealed a multiplicity of binding states for adsorbed atoms showing why the initial adsorption on clean surfaces occurs at specific sites, areas, or planes of the surface.

Low-energy electron diffraction (LEED) is employed to study two-dimensional surface crystallography [60]. Since X-rays and high-energy electrons penetrate the bulk of the crystal unless grazing angles of incidence are used, such energetic beams are normally reserved for determining the three-dimensional structure of crystals. However, electrons with an energy of the order of 100 eV are predominantly scattered by the first layer of atoms on a crystal and their diffraction patterns may be used to study two-dimensional structures [61] on solid surfaces [62]. A diffuse pattern indicates a microcrystalline or noncrystalline surface layer. During recent years, experimental techniques have become sufficiently developed and improved to make low-energy electron diffraction an effective research tool [63]. Somorjai [18,27] has collected experimental results up to approximately 1979. He includes important information on mechanisms and structural relationships found in the chemisorption process and during initial oxide formation. These are discussed below.

Low-energy electron diffraction studies have established that adsorbed gases may form a variety of structures on solid surfaces depending on the adsorbent and the adsorbate, exposed crystal face, coverage, temperature, impurity atoms, and defects. The adsorbed layer may range from disordered, like that found in adsorption of oxygen on silicon or germanium, to regularly ordered structures of adsorbed gas atoms which do not cause any detectable effects on the substrate, as for instance, oxygen on the (100) face of nickel. The adsorption of gases may also be accompanied by a rearrangement of the surface atoms of the substrate, as observed for oxygen adsorption on the (110) nickel surface.

Secondary ion mass spectroscopy (SIMS) and electron-induced desorption (EID) are two techniques used to identify adsorbed species on a metal or oxide surface [53]. The former technique utilizes a beam of gas to sputter away the surface of a specimen. A mass spectrometer located near the point of impact is then used to identify the sputtered material. The matrix effects in SIMS enhance the sensitivity of oxide surface analysis compared to a metal surface. The EID technique uses electrons to desorb ions from a surface. This approach complements the use of electron-spin resonance in the identification of adsorbed species. It should be noted however that the energetic beams of ions or electrons can change the electronic state of the adsorbed species so that the original state of adsorption may not be observed.

The nomenclature of surface phases has been codified over the years [61,64] so that the various structures and order–disorder phenomena may be specified. As yet, no extensive phase diagrams of two-dimensional phases are available.

A convenient technique for comparing rates of reaction under different sets of conditions, but one difficult to interpret directly in terms of oxide growth kinetics, is change in electrical resistivity of a metal film upon exposure to oxygen. Hunt and Ritchie [65] among others have used the method with some success. They applied the equation

$$x = \frac{\sigma' \rho l}{w}\left(\frac{1}{R_0} - \frac{1}{R}\right) \tag{1}$$

where film resistance $R = k'f(x)P^n$

x is the oxide thickness

σ' is the metal–oxide thickness conversion factor

ρ is the metal film resistivity

l is the film length between contacts

w is the width of film

R_0 is the initial film resistance

$k'f(x)$ is the oxidation rate

P is the oxygen pressure

n is a constant

to the oxidation of aluminum. The exponent n of pressure dependence

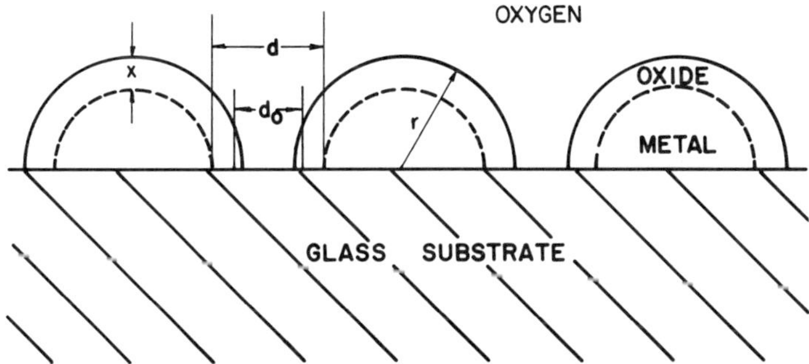

Figure 5.1. Idealized cross section of an oxidized discontinuous metal film. x is oxide thickness, d_0 is initial interisland distance, r is island radius, and d is metal island separation after oxidation. After Fehlner [67]. Reprinted by permission of the publisher, The Electrochemical Society, Inc.

for the reaction decreased from an initial value of $\frac{1}{2}$ with both increasing oxide thickness and temperature. Interpretation of kinetics by this technique is limited by difficulties in determining ρ and R_0. The use of Equation 1 for chromium and the effect on the oxidation process of various parameter changes is reviewed by Hope and Ritchie [66].

A more sensitive resistivity technique for studying low-temperature oxidation is based on the use of discontinuous metal films deposited on an insulating substrate [67,68,69,70]. The method complements single crystal studies and is closely related to the theory of electrical conductivity in cermet films [71]. An idealized film is shown in cross section in Figure 5.1. It consists of metal hemispheres of radius r separated from each other by distance d. Generally, both dimensions are in the range 10–100 Å. Upon oxidation, the metal is converted to an oxide, thereby increasing the separation between the residual metal islands. Since electronic conductivity in such films occurs by tunneling from island to island via the substrate, conductivity varies exponentially with the separation between the islands.

Neugebauer and Webb [72] have analyzed electron transport in discontinuous metal films. The process of carrier generation is thermally activated. It takes work approximately equal to $e^2/\kappa r$, where κ is the dielectric constant of the transport medium, to remove an electron from one island to an adjacent island. In the absence of a field, charge-carrier generation and recombination take place in the thermally activated equilibrium process.

A net movement of charge carriers can occur in a particular direction if a field is applied parallel to the film. Electron transfer from island to island occurs by tunneling via the substrate rather than through the surrounding gas or vacuum because the substrate dielectric constant is higher. The overall process is described in an expression for the resistivity ρ of a discontinuous metal film.

$$\frac{1}{\rho} = B_1 d\phi^{1/2} \exp\left(-\frac{W''}{kT}\right) \exp(-B_2 d\phi^{1/2}) \tag{2}$$

where ϕ is the potential barrier to electron transfer, W'' is the activation energy for electron transfer, equal to $e^2/\kappa r - e^2/[\kappa(d + r)]$, and B_1, B_2 are constants if r is assumed constant.

During oxidation, the interisland distance d increases by twice the oxide thickness x so that the tunneling electrons must traverse a longer distance. There is, of course, a distribution of island sizes and interisland distances in any real film, but this refinement in the mathematical treatment has not yet been included in the equations.

Any change in r and thus W'' has also been ignored because of the low temperatures involved and the fact that for any change in r there is twice the change in d, especially where r is large and d is small. The value of ϕ appears fixed since tunneling occurs at the metal–substrate interface.

If we differentiate Equation 2 with respect to d after taking the logarithm of both sides, we obtain

$$-\Delta \log \rho = \Delta \log d \left[1 - \frac{B'' \phi^{1/2} d}{\hbar}\right] \tag{3}$$

where B'' is a constant and $\hbar$ is $h/2\pi$. Thus, we see that a change in d due to oxidation is reflected by a change in film resistance R since $R = \rho l/\Lambda$ where l is the distance between contacts and Λ is the cross-sectional area of the conductor. A plot of log R versus d should give a straight line.

Empirically, growth of three-dimensional oxide of thickness x at low temperatures is often expressed as a direct logarithmic function

$$x = k' \log(a't + t_0) \tag{4}$$

where a' is a constant and the constant t_0 can be neglected for large t. Since $d = d_0 + 2x$ such that $d = d_0 + 2k' \log t + 2k' \log a'$, a plot of log R versus log t should yield a straight line from which k', the direct logarithmic rate constant, can be determined at any temperature and oxygen pressure. Data showing a linear relationship for titanium–oxygen are shown in Figure 5.2 [68].

A similar relationship can be worked out for inverse logarithmic kinetics where

$$\frac{1}{x} = k''(a'' + \log t) \tag{5}$$

Equation 3 becomes

$$\log \rho \cong \text{const} - \frac{8\pi(2m'\phi)^{1/2}}{2.3\hbar k'' \log t} \tag{6}$$

when $\log t > a''$. A plot of $\log R$ versus $1/\log t$ provides a way to determine k''. When the data from Figure 5.2 are plotted in this way, chemisorption and oxidation regions can again be identified. See Figure 5.3.

Rutherford back scattering (RBS or HEIS) can be used to characterize oxide films. It is based on analyzing the ion energies of reflected high-energy ions which have bombarded a surface [73]. Inelastic scattering events in the solid yield ion energies characteristic of the target atom and its location at or near the surface of the solid. Both compositions and profiles can be determined from this data.

5.2.3. Oxygen Adsorption

Adsorption is the initial step in the reaction between a clean metal surface and a gas. When this initial part of the process involves solely chemisorption on the clean surface, the sticking coefficient has been found to be close to 1 [74,75] for metals such as W, Mo, Ti, V, Zr, Hf, Nb, Ta, Ni, and Fe. Others such as Be [76], Zn [77], Al [78], Au, and Si [79] have smaller values, of the order of 0.1 [80] or less.

The sticking coefficient can be dependent on both temperature [81] and crystallographic orientation [82] as seen in the case of tungsten. Steps and kink sites on the metal surface are of great significance during chemisorption [83,84,85]. Electron bombardment can change the rate of adsorption [86].

When a monolayer or a fraction of a monolayer of oxygen has been formed, depending on the two-dimensional structure, an abrupt decrease in the sticking coefficient is observed. This can be masked, however, by a reaction rate which is linear with time and proportional to oxygen pressure. In this case, a constant value of sticking coefficient is found.

The behavior of two-dimensional matter [87] has gained a great deal

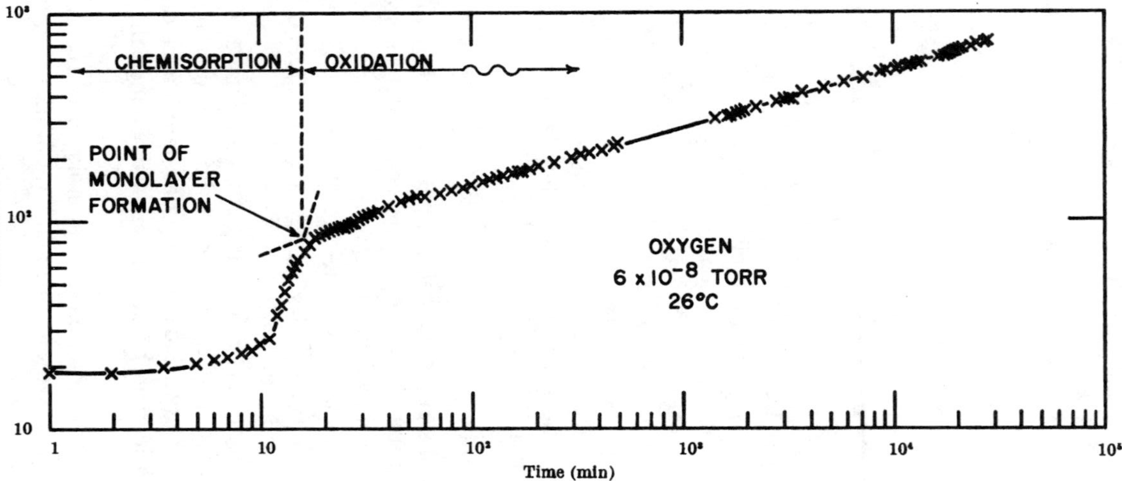

Figure 5.2. Log–log plot of the resistance of a discontinuous film of titanium as a function of time during exposure to oxygen. After Fehlner [68]. Reprinted by permission from *Nature*, Vol. 210, No. 5040, p. 1035, Copyright 1966, Macmillan Journals Limited.

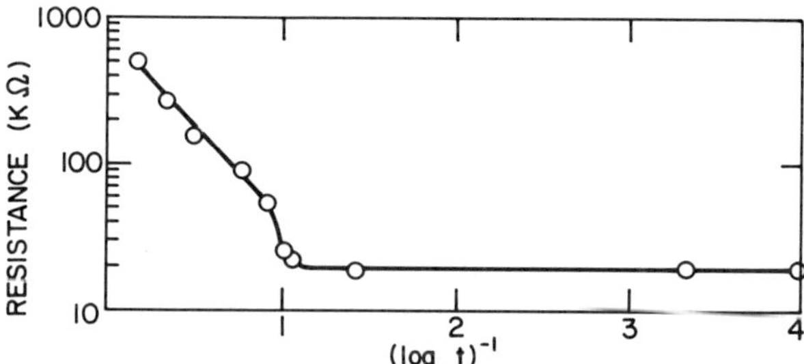

Figure 5.3. Plot of the log of the same resistance data as shown in Figure 5.2 as a function of 1/log *t*.

of attention during the past three decades. Conclusions of Somorjai [18] based on LEED results are summarized here. The set of empirical rules which govern the formation of surface structures for chemisorbed gases on high-density crystal planes may be stated as follows:

1. The adsorbed species prefer close-packed arrangements in which adsorbate–adsorbate and adsorbate–substrate interactions are maximized. Structures of adsorbed species dominate in which the unit cell is equal to or up to twice as large as that of the substrate.
2. The ordered structures formed by the adsorbed species usually have the same rotational symmetry as the substrate.
3. The structure of a one-to-two monolayer thick surface layer often resembles that of the substrate rather than that of the bulk condensate.

Small atoms such as oxygen or sulfur occupy high-symmetry surface sites at atomic distances characteristic of a covalent bond to the nearest neighbor metal atoms. The presence of surface irregularities, steps, and kink sites at a surface can lead to different bonding characteristics when compared with atoms on surface terraces. For instance, Collins and Spicer [88] attribute a higher sticking probability for dissociative oxygen adsorption on Pt steps or defects than on Pt (111) terraces.

These guidelines may be applied to the case of oxygen chemisorption. It is assumed here that the temperature is low enough that oxygen does not dissolve to any great extent in the bulk metal, but instead, remains as a partially ionized surface layer [89]. Incoming oxygen molecules which hit a clean metal surface can adsorb on it, subsequently dissociating at adjacent sites [90,91]. The activation energy for the overall process is very low since the process occurs even at liquid helium temperature [6,92]. During this period, the sticking coefficient α remains high. It may vary as $(1 - \theta)$ for single site adsorption or as $(1 - \theta)^2$ for dissociative adsorption on a single crystal surface for coverage $\theta < 0.25$. A polycrystalline surface with its variety of crystal faces plus steps and kink sites would obscure such detail. In this case, α would remain high until a stable oxide layer had formed.

To account for sticking coefficients smaller than 1, Ehrlich [4,21] and Farnsworth et al. [93] have proposed that the gas molecules first become physically adsorbed on the surface and that subsequent chemisorption occurs at specific sites on the surface from the precursor state of a reservoir of physically adsorbed molecules. Redondo et al. [94] have calculated that O_2 can bind to silicon at a single surface atom, giving –Si–O–O. A precursor state is also postulated for oxygen adsorption on Ge (100) [95].

The heat of adsorption ΔH remains high during initial chemisorption [96], close to the value for the heat of formation of the oxide [2,27].

A metal single crystal shows a complex α versus θ behavior. Holloway and Hudson [97] have shown that the initial high sticking coefficient first decreases and then increases before monolayer formation, as shown in Figure 5.4. They interpret this behavior in terms of a two-dimensional equilibrium between chemisorbed oxygen and island formation as discussed below.

Linear kinetics of the initial process occurring during low-temperature oxidation has been expressed by Fehlner and Mott [2] and others [98] as

$$\frac{dx}{dt} = k_0 P \tag{7}$$

where x is the oxide thickness, P is the oxygen pressure, and k_0 is a constant. Copper has been found to follow this equation [99]. Another example is the work of Unertl and Blakely [100] on the oxidation of Zn (0001). During the initial stages, oxide growth was linear in pressure and independent of temperature from -196 to 152°C. Extensive data on the electronic and atomic structure of the oxide are reported for this process. Below room temperature, the oxide was either noncrystalline or microcrystalline (LEED) while above, it was crystalline. Heterogeneous growth of the oxide with no distinguishable precursor adsorbed state is proposed. Oxide growth proceeded in this fashion until the metal surface was covered to a mean thickness of two to three monolayers.

Fowler and Blakely [101] studied the initial stages of Be (0001) oxidation and found it to be linearly dependent on oxygen pressure at high pressures. Strong enhancement of oxidation by the AES electron beam interfered with determining intrinsic oxidation rates. See Chapter 8.

A distinction between network-forming and modifying oxides can be

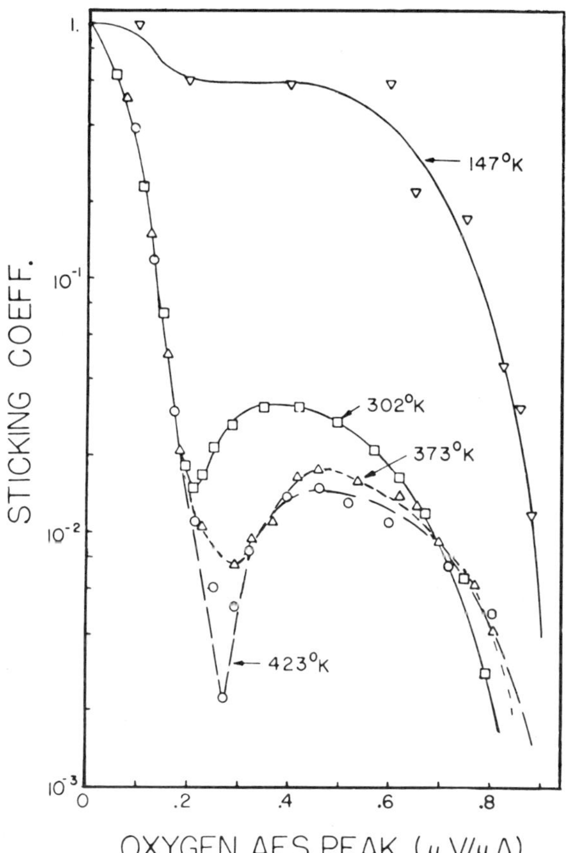

Figure 5.4. Oxygen sticking coefficient versus oxygen coverage on Ni (100) as a function of temperature. After Holloway and Hudson [97]. Published by permission of the copyright holder, North-Holland Physics Publishing.

observed at room temperature, even in monolayers of oxygen. For metals whose oxides form networks, the LEED patterns are often disordered, depending on the crystal face being studied. Examples are aluminum, silicon [18], and chromium [102]. As_2O_3 formed by oxygen adsorption on arsenic was studied by photoemission and shown to be disordered [103].

On the other hand, metals whose oxides are modifiers tend to produce well-defined LEED patterns characteristic of crystalline oxide. Epitaxial relationships can even be observed, as with oxides on nickel [104] or iron [98]. It appears that bonding in the network-forming oxides is so strong that rearrangement to form a surface structure having long-range order is often impossible at room temperature. Network modifiers, on the other hand, appear able to rearrange to form crystals at room temperature.

The initial oxygen adsorbed on a metal surface tends to occupy every other adsorption site, due to repulsion between charged oxygen ions [105]. This charge is due to partial electron transfer from the metal to the oxygen atoms so that the metal–oxygen bond has a partially ionic, partially covalent character.

When approximately half a monolayer of oxygen has formed, only single cation sites remain for adsorption of oxygen molecules. There is discussion in the literature concerning various mechanisms for continued reaction. Molecules may be adsorbed at single cation sites [106], double bonded oxygen atoms may form [107], or surface rearrangement (place exchange) may make available adjacent cation sites for continued dissociative adsorption of oxygen [93,108]. A combination of mechanisms may actually occur.

Krueger and Pollack [109] have treated this point in more detail. They distinguish normal chemisorption from chemisorption with incorporation. In the former, the number of adsorption sites is fixed so that the sticking coefficient is a function of coverage. In the latter, place exchange of the initially adsorbed oxygen makes new adsorption sites available so the number of sites remains essentially constant. Models of oxygen adsorption must satisfy both oxygen uptake versus time and the rate of uptake versus total uptake, for example, α versus θ and $d\theta/dt$ versus θ. Krueger and Pollack interpret results for initial aluminum oxidation in terms of chemisorption with incorporation followed by normal chemisorption of the dissociative type. This results in the formation of a stable oxide film greater than a monolayer in thickness. Carley and Roberts [110] suggest an incorporated oxygen structure which is a precursor to formation of Al_2O_3 on aluminum.

The study of initial oxide formation has been complicated owing to difficulties in producing clean surfaces and the stringent high-vacuum requirements needed for maintaining them [111]. Many studies of the oxidation of metals have been made on metals whose surfaces are already covered with a thin oxide film and/or impurities at the beginning of the experiment. For instance, carbon on nickel does not prevent oxide growth [112] but interferes with the rate of initial oxide formation [113]. Windawi [114] proposes using sulfur as a marker to study the transition from chemisorption to oxide growth on nickel. The details of carbon and sulfur segregation to a nickel surface have been examined by Mróz et al. [115]. Klein et al. [116] report that sulfur on Ru $(10\bar{1}0)$ prevents the adsorption of oxygen. Graphitic carbon on silicon can be converted to a carbide during oxidation and remain at the Si/SiO_2 interface [117]. It appears then that studies of initial oxide formation on surfaces known to be clean are of recent date, and as yet only a limited number of such studies have been reported.

As discussed above, the initial step in the reaction between a metal and oxygen is either chemisorption or adsorption–incorporation of gas on the metal surface. MacRae et al. [118,119] in their low-energy diffraction studies of the interaction of oxygen with clean nickel surfaces were able to study the initial formation of three-dimensional oxide taking place after the surfaces had become fully covered with chemisorbed oxygen. An important feature of this initial oxide formation was that isolated oxide nuclei formed at what appeared to be random positions on the surface. The same observations have been made for initial oxide nucleation on other metals as discussed below. The nature of the nucleation sites is still a matter of conjecture, but they may possibly constitute surface imperfections [120], impurity atoms, or both. After formation of the oxide nuclei, oxidation can proceed through lateral growth of the individual crystallites until the whole surface is covered with oxide.

5.3. EXPERIMENTAL EVIDENCE CONCERNING INITIAL OXIDE FORMATION

A concise illustration of the transition region and formation of a stable oxide film is found in the work of Kirk and Huber [121]. Aluminum oxidation was studied under dry conditions of 25°C and various oxygen pressures, as shown in Figure 5.5, where film resistance, weight gain, and surface potential are shown as a function of oxygen exposure. The transition from linear to logarithmic kinetics occurs in the vicinity of 100 L where L is 1 Langmuir equal to 10^{-6} Torr sec. Approximately the same transition point is shown by each measurement technique [122]. Interestingly enough, most metals at low temperatures exhibit a similar transition in the same range of oxygen exposure when the effect of impurities has been eliminated [2], an example being Ni [26].

The transition from chemisorption to the growth of three-dimensional oxide is illustrated by Figure 5.6 from the work of Mitchell et al. [123] on Ni (110) and Ni (001). Reflection high-energy electron diffraction (RHEED) and X ray emission were utilized to study initial processes during the interaction of oxygen with nickel (40–300°C). Cleaning of the surface involved oxidation, argon ion bombardment, and hydrogen reduction to avoid rate retardation by impurities such as carbon. Three stages of oxidation were identified: chemisorption, island growth, and three-dimensional oxide growth. These three stages were also identified by Holloway [124] for nickel oxidation.

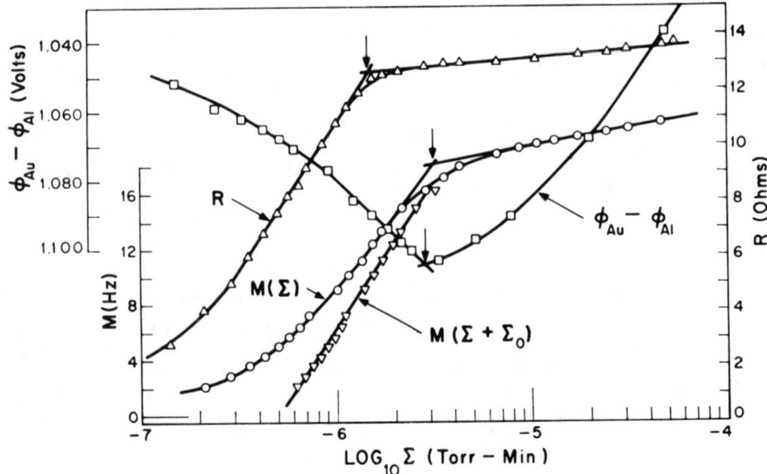

Figure 5.5. Change of film resistance R, microbalance reading M, and contact potential difference $\phi_{Au} - \phi_{Al}$ versus oxygen exposure in Torr min. The original data M (Σ) was modified by $\Sigma_0 = 4 \times 10^{-7}$ Torr min to linearize the data. Microbalance sensitivity was 0.28 Å of bulk Al_2O_3 Hz^{-1}. The vertical arrows mark the transition from fast initial oxidation to slow logarithmic oxidation which occurred at exposures between 60 and 210 L. After Kirk and Huber [121]. Published by permission of the copyright holder, North-Holland Physics Publishing.

Benninghoven et al. [125], in an extensive study of surface reactions using SIMS, AES, EID, and flash filament techniques, have discussed these three stages in terms of chemisorption (perhaps activated, with dissociation into atoms), transition to three-dimensional oxide, and further oxide growth, perhaps with dissolution of oxygen in the metallic substrate. The activated chemisorption may be related to the molecular precursor reported by Leung and Gomer [6]. It has also been proposed by Wagner [126] as the rate-limiting step during the initial oxidation of nickel. Roberts [127] has interpreted iron oxidation subsequent to the fast uptake of oxygen at -195 and -80°C as due to surface diffusion followed by chemisorption of molecularly adsorbed oxygen. Weinberg et al. [128] propose a mobile precursor state during chemisorption of oxygen on Ir (111).

For the case of nickel [123], the three stages may be summarized as follows.

5.3.1. Stage 1—Chemisorption

Directly proportional to pressure, independent of temperature; sticking coefficient equal to 1; continuing to half a monolayer of NiO on the (110)

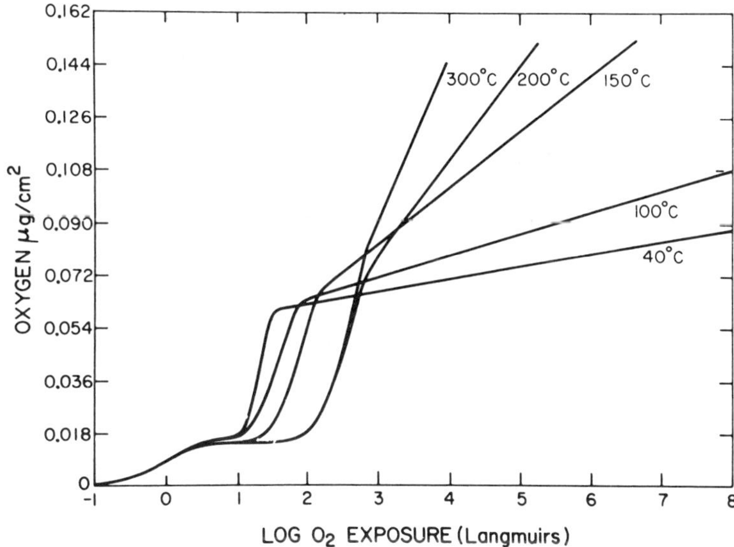

Figure 5.6. Oxygen uptake by Ni (110) as a function of oxygen exposure at various temperatures showing three stages of oxidation: chemisorption, island growth, and slow logarithmic growth. 1 μg cm^{-2} O_2 is equivalent to 69.2 Å of bulk NiO. After Mitchell et al. [123]. Published by permission of the copyright holder, North-Holland Physics Publishing.

face while on the (001) face, the limit is a third of a monolayer; structures progressing from (2 × 1) through (3 × 1) to (9 × 4) on the (110) face while on the (001) face, they progress from *p* (2 × 2) to *c* (2 × 2). Helium ion backscattering [129] shows that the oxygen protrudes from the nickel surface by 0.4 to 0.8 Å.

5.3.2. Stage 2—Oxide Island Growth

Directly proportional to pressure but inversely proportional to temperature; continues to a coverage of two monolayers for (110) and three for (001); NiO grows in (001) epitaxy on the Ni (110) substrate and is compressed 4.5% in the $[1\bar{1}0]$ direction on the plane of the surface; two epitaxial relationships, (001) and (111) oxide, exist for Ni (001); nucleation and growth follow the equation of Holloway and Hudson [97] based on a fixed number of nucleation sites and lateral growth from the island perimeter.

5.3.3. Stage 3—Logarithmic-Type Growth

Directly proportional to pressure and temperature; NiO islands coalesce and a normal $[1\bar{1}0]$ lattice spacing develops; the simple (001) epitaxial relationship is replaced by a more complex one of the $(1\bar{1}7)$ type; direct logarithmic kinetics have an activation energy of 0.25 eV and are explained by electron tunneling. Evans et al. [130] discuss the oxidation of nickel in terms of dissociative chemisorption followed by monolayer formation and penetration of oxygen below the metal surface. A nonuniform oxide forms composed of nonstoichiometric nickel oxide.

Further examples illustrating these stages of initial oxide growth may be found in the literature on oxidation [131]. The Vienna Conference on Surfaces [52] contains reports on studies of Al, Cu, Fe, W, Ti, Pt, Ir, Ni, Zn, Co, Mo, Pd, Si, Ag, Sb, and Pb. The chemisorption stage is emphasized. Iron oxidation has been extensively examined in a molecular beam study by Dorfeld et al. [132]. Initial chemisorption was succeeded by growth of an oxide interpreted to be FeO. The first two stages during the oxidation of chromium, chemisorption and island formation, have been observed using LEED and AES [102]. Work function measurements led Heyne and Tompkins [133] to postulate chemisorption, incorporation, nucleation, and finally growth of La_2O_3 on lanthanum at 27°C.

Nyberg [134] prefers to divide the initial oxidation of Ca, Sr, and Ba into a low- and high-exposure regime. The exposure for oxide formation was determined to be 70 L for Ca, 110 L for Sr, and 480 L for Ba.

Sakisaka et al. [135] extend the number of stages to four for Cr: adsorption, incorporation, rapid oxidation to produce Cr_2O_3, and slow thickening of the oxide film. The oxidation of bismuth [136] showed the three stages of growth discussed above, as did nickel [123]. Copper [137] is reported to give chemisorption, Cu_2O island formation, and a thin film of bulk cuprous oxide. Titanium [138] appeared to have two or four stages depending upon the analytical technique used.

The existence of oxide with an intermediate oxidation state has been reported using PES in a study of the initial oxidation of aluminum [139]. Bujor et al. [140] using AES, ELS, and ESD did not confirm this intermediate state. Gallium [141] possibly shows an intermediate oxidation state. Two states have been found for nickel [130] where Ni^{2+} and Ni^{3+} are proposed. In the case of chromium [142], a spinel-like oxide converts to rhombohedral Cr_2O_3. For iron [143], the Fe^{2+}/Fe^{3+} ratio decreases with increasing pressure. Oxygen produces CoO on cobalt at 22°C and 5×10^{-8} Torr but Co_3O_4 at a pressure of 0.25 Torr [144]. Boron showed no evidence of an intermediate oxidation state [145]. Silicon has four oxidation states [146] corresponding to silicon atoms bonded to one, two, three, or four oxygens. Bonding geometries differed on the Si (111) compared to Si (100). Lieske and Hezel [147] report the formation of various silicon–oxygen bonds: Si=O and Si–O of the type found in $SiO_{4/2}$ tetrahedra, Si–Si bonds, and broken Si–O bonds.

Tantalum [148] oxidation below 500°C proceeds through the precipitation of a suboxide with the subsequent formation of "amorphous" Ta_2O_5. Niobium [149] oxidation sometimes showed NbO in addition to Nb_2O_5. Plutonium [150] also showed a suboxide in addition to PuO_2. Copper [151] forms Cu_2O.

The lanthanide series has been examined for the members terbium to lutetium [152]. A logarithmic growth rate was found for all members except ytterbium which was linear. Island growth during reaction with water is postulated. In the case of yttrium [153], an intermediate state was found which consisted of both oxidized and unoxidized yttrium atoms. Oxygen exposure of Sn, Gd, and Tb has been studied by AES [154]. The initial oxidation of zinc and cadmium [155,156] was interpreted in terms of island growth. Surface reconstruction on Cu (110) accompanied by island growth is reported by Lapujoulade et al. [157]. Some noble metals adsorb little or no oxygen, for example, Pd (111) [158].

5.4. TRANSITION PERIOD—OXIDE ISLAND GROWTH

When a molecule of oxygen adsorbs on a cation site, it can dissociate if an adjacent cation site is already available or is made available by some process such as place exchange [2,159]. In this process, shown in Figure 4.4, an underlying metal ion changes places with an adsorbed oxygen ion, preferably at a kink site or grain boundary. Since no oxide structure exists at this point in the process, it is inappropriate to speak of individual anion or cation movement. Instead, it is proposed that the anion is incorporated by a rearrangement of the surface metal atoms and both species end up moving. Additional oxygens may then dissociate by forming bonds with the adjacent cations. In this manner, a second layer of oxide begins to form before the first one is complete, marking the beginning of the transition from a two-dimensional monolayer to a three-dimensional oxide. During this process, oxide islands can nucleate, grow, and coalesce into a continuous oxide.

The average sticking coefficient and heat of adsorption during this period remain high on metals, but decrease precipitously upon formation of a stable oxide layer. The formation of such a layer at low temperatures initiates a transition from the linear kinetics of chemisorption and place exchange to logarithmic kinetics. At higher temperatures the transition can be to parabolic or a modified linear kinetics. In the case of inverse logarithmic (Cabrera–Mott) kinetics, an oxide thickness may be defined where dx/dt in Equations 21 and 43 of Chapter 2 are equal [160]. The resulting oxide thickness at which the transition occurs is

$$x = \frac{qaV}{W - 2.3\,kT\log(N\Omega\nu/k_0P)} \tag{8}$$

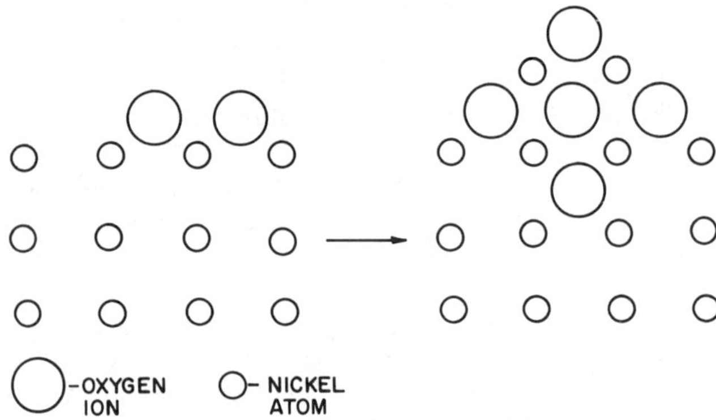

Figure 5.7. Diagram of the transition from adsorbed oxygen on nickel to three-dimensional islands of NiO.

where the variables are defined in Chapter 2. The transition is complex and its details depend upon the metal being studied, temperature, pressure, surface perfection, and cleanliness.

This transition from chemisorption to place exchange marks the beginning of three-dimensional oxide growth as illustrated in Figure 5.7. It is apparent that the formation of a second and third layer of oxide before the first one is complete will produce an island structure on the scale of a few tens of Angstroms.

5.4.1. Types of Islands

Two types of oxide islands must be distinguished. The first is statistical in nature and the adsorbed species are assumed to be immobile. Hence, it may occur for oxidation at cryogenic temperatures. Before the first layer of atoms is completely formed, a second layer begins to grow. For example, a statistical study of tungsten island growth from the vapor phase [161] has shown that up to four tungsten layers can be growing simultaneously before the first layer is complete at temperatures below $-196°C$.

The second type of island depends upon surface diffusion for growth. It nucleates either homogeneously on metal terraces, or heterogeneously at ledge and kink sites on the metal surface, or at impurities. It then grows laterally and/or vertically by surface diffusion of the adsorbed species.

Island features of oxide growth were highlighted by Bardolle and Bénard [162] for oxidation of iron at moderate temperatures. Later studies have revealed that growth of discrete oxide particles appears to be a general phenomenon during initial stages of the oxidation of metals. Lu

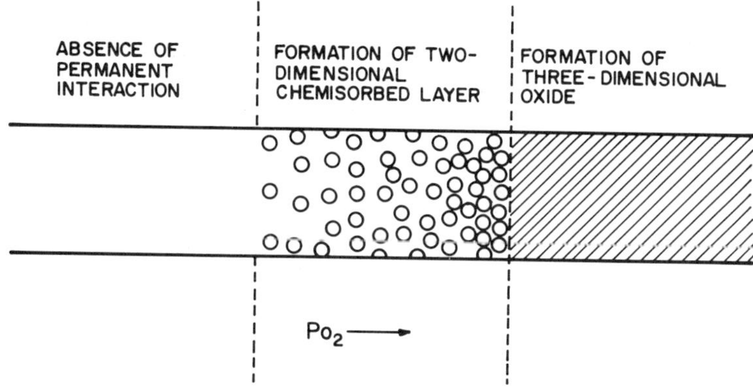

Figure 5.8. Effect of oxygen partial pressure on the nature of the surface structure for oxygen on a metal. From J. Bénard, "Adsorption of Oxidant and Oxide Nucleation," *Oxidation of Metals and Alloys*, American Society for Metals, 1971, p. 2.

et al. [163] present a model in which condensation of islands having ordered structures occurs from a chemisorbed overlayer.

This process has been summarized by Bénard et al. [164,165]. They have divided the interaction of a metal surface into three zones based on oxygen pressure: (A) where the clean metal is stable, (B) where a chemisorbed monolayer is the equilibrium state, and (C) where oxide islands can nucleate and grow to form three-dimensional oxide as shown in Figure 5.8. Island growth according to the second type described above is observed experimentally if two conditions are met: the temperature is sufficient to allow surface diffusion to occur and the reaction takes place slowly under conditions of limited oxidant concentration. This is shown for copper at 550°C in Figure 5.9 [166].

Bardolle and Bénard [167] have also shown that crystal orientation of the substrate metal affects the number of nuclei formed on iron. They are more numerous on the (100) than on the (110) face by a factor of 100 to 1.

Summaries covering island growth have been written by Bénard [168], Gwathmey and Lawless [169], and Fischmeister [170]. A similar process has been observed during the sulfidation of copper [171]. Ritchie [172] and Oudar [173] treat the process of adsorption both with and without oxide nucleation. The effect of surface defects on island nucleation is sometimes observed at low temperatures and pressures but not at high temperatures and pressures where homogeneous nucleation can take place. It seems plausible that a layer of adsorbed oxygen or even a thin oxide film can exist on the metal between the oxide islands [124,174].

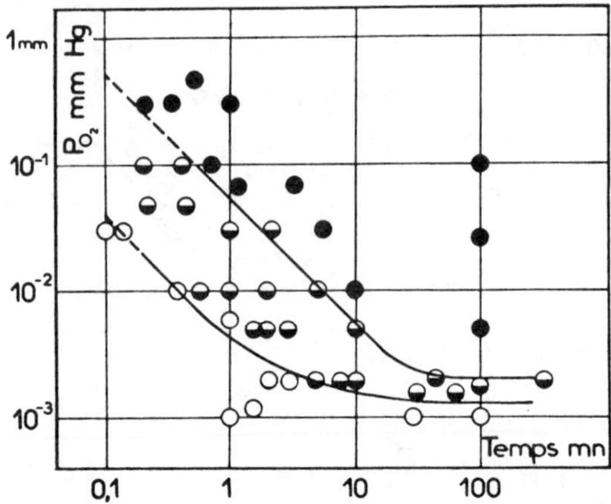

Figure 5.9. Region of nuclei formation on a crystal of copper oxidized at 550°C as a function of oxygen pressure (mm Hg) and time (mn). Solid circles represent continuous film, half circles oxide nuclei, and open circles bright metal. After Grønlund [166].

5.4.2. Mechanism of Island Growth

Island formation has been proposed to explain the oxidation of barium films produced by evaporation in high vacuum [175,176,177,178] and clean evaporated magnesium surfaces as studied by Cohen [179] and Adiss [180]. For both metals, the reaction with oxygen was studied using the Wagener method [175] in which the oxidation rate is followed by measuring the rate of oxygen flow to the metal surface through a tube of known conductance.

It was found that the sticking coefficient for oxygen during the initial stage of reaction increased with time. Bloomer [178] proposed that this increasing activity is due to an initial oxide nucleation at preferred sites and that the reaction proceeds through a lateral surface growth of the nuclei. It is assumed in the model that the surface is heterogeneous and that oxide nucleation occurs at certain sites while the rest of the surface is comparatively inert to attack. In this connection, it may be noted that Bloomer also observed an initial induction period during which no oxygen was consumed and thus no oxide was nucleated. This observation may relate to the surface perfection of the evaporated barium films. In like manner, the corrosion resistance of glassy metals leads one to believe that nucleation may be difficult on the atomically smooth surface of a glassy metal [181] as pointed out in Chapter 3.

If it is assumed that only the surface or edge of oxide nuclei (islands) are active in the oxidation, the islands will grow laterally and the sticking probability will increase with the extent of oxidation. Thus, the sticking coefficient α as a function of oxygen uptake z can be expressed by the equation

$$\alpha = A_6 z^{1/n} \tag{9}$$

where A_6 and n are constants.

The values of n will be determined by the geometry of the oxide islands and the rate-determining step in the process. Bloomer considered two cases: (1) The oxide islands are cylindrical disks which are one monolayer high, and the active region for oxygen uptake is an annulus one atom wide around the circumference of the island. The value of n is 2 if only the oxygen atoms impinging within the annulus are taken up. (2) The islands are hemispherical caps and the whole oxide surface is active in oxygen uptake. In this case n is $\frac{3}{2}$. Orr [182] found experimentally that α varies as $z^{1/2}$ and he concluded on this basis that magnesium oxide nuclei behave as two-dimensional disks with a narrow reaction zone at the edge. Oxide island density was estimated by Orr to be 2×10^{11} cm^{-2} with a mean separation distance of 200 Å.

A possible model for the island growth process can be described as follows. Chemisorbed oxygen ions on a surface initially repel each other. This array of dipoles can lower its total energy through polarity reversal of alternate dipoles. Only when a cation moves in between two oxygen ions does a lateral attraction occur. Such movement by place exchange, shown in Figures 4.4 and 5.7, takes place most easily at a step or defect on the metal surface. The presence of adsorbed oxygen on the chemisorbed monolayer also facilitates the growth of an island. The migration of oxygen over the chemisorbed layer to the site of place exchange increases the rate of place exchange and thus the rate of island growth. The island growth may be envisaged as an advancing wave of place-exchanging ions, fed by lateral diffusion of oxygen over both the chemisorbed layer and the surface of the growing oxide. The incorporation of oxygen at the periphery of the growing island is what keeps the sticking coefficient high in the case of polycrystalline films and explains the complicated change in α with θ for single crystal surfaces, as shown in Figure 5.4.

Lee [160] points out a related model which stresses the role of low-work function centers such as steps and ledges. He proposes that on a high-work function metal surface, favorable sites for oxide nuclei are provided by faceting of the metal surface. If this occurs, then each oxide island would be located over a shallow pit in the metal surface.

After the period of increasing sticking probability, the studies on both barium and magnesium showed an abrupt decrease in the rate of oxida-

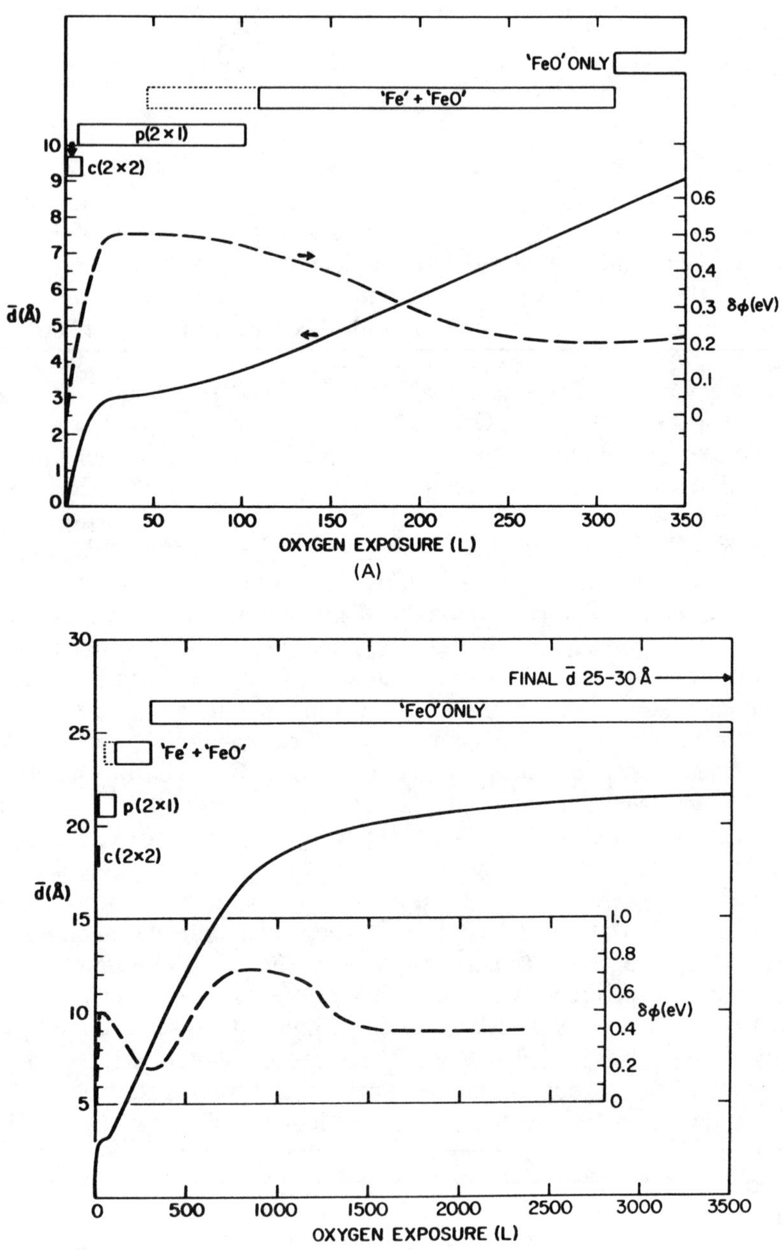

'FeO' ONLY
'Fe' + 'FeO'
p(2×1)
c(2×2)
d̄(Å)
δφ(eV)
OXYGEN EXPOSURE (L)
(A)
FINAL d̄ 25–30 Å
'FeO' ONLY
'Fe' + 'FeO'
p(2×1)
c(2×2)
d̄(Å)
δφ(eV)
OXYGEN EXPOSURE (L)
(B)

tion. This change has been interpreted to mean that the islands have grown together to cover the entire metal surface. For magnesium, Orr [182] found the abrupt decrease occurred at an oxygen uptake equivalent to 3–4 monolayers of oxygen, that is, the oxide islands appear to be about 3–4 monolayers high. Crystallite size was estimated to be about 100 Å, a value in good agreement with the nucleation density given above.

5.4.3. Experimental Evidence for Island Growth

Another example of island formation at low temperatures is found in the work of Melmed and Carroll [183]. They performed a combined study of LEED, ellipsometry, and field emission measurements during the dry oxidation of single-crystal iron films at 25°C. The iron films were epitaxially grown on the (110) face of a tungsten crystal. The LEED structure, thickness, and work function of the oxide film are shown in Figure 5.10 as a function of oxygen exposure. The initial period of linear oxidation (chemisorption and place exchange) is readily apparent. The presence of island growth during the transition period is deduced from two observations. First, a maximum occurred in the work function followed by a minimum, as also observed by Hall and Mee [184]. The minimum is presumably due to geometric enhancement of the field. Secondly, LEED patterns interpreted to be Fe and FeO were simultaneously observed. The subsequent logarithmic growth period was characterized by a constant surface potential and the presence of oxide on the sample surface.

No difference in behavior between network-forming and modifying oxides has been experimentally detected for island growth. However, it is expected that island size should differ, being smaller for network formers.

Holloway and Hudson [97] propose that for nickel the number of growing islands is directly related to the number of defects on the clean metal surface. Eventually all the islands grow together, producing a continuous oxide film. When the defect density is very high as in polycrystalline films, then island coalescence would occur too fast to be measured experimentally. The sensitivity of oxygen adsorption on copper to surface micro-

←

Figure 5.10. LEED, ellipsometry, and work function results for oxygen on (011) iron at room temperature. (A) Average film thickness $\bar{d}$ (solid line) derived from ellipsometry data, electron work function change $\delta\phi$ (broken line) derived from surface potential data, and LEED patterns versus oxygen exposure in Langmuirs. Vertical arrow indicates maximum intensity of C (2 × 2) pattern. Dashed line extension of 'Fe' + 'FeO' exposure range indicates extension of that range inferred from ellipsometric and surface potential data. (B) Same type of display over 10 times greater exposure range. After Melmed and Carroll [183]. Published by permission of the copyright holder, the American Institute of Physics.

structure has been emphasized by McKee et al. [185]. Thomas [186] comments on the disagreement over the role of dislocations in oxide island nucleation. He points out the synergism between impurities and dislocations in promoting nucleation. Barna [187] discusses the accumulation of impurities at atomic steps which represent the most reactive sites on a surface. The impurities in turn can interfere with continued reaction.

Observation of oxide islands requires a technique such as field ion microscopy [59], discussed in work by Brenner and McVeagh [188], Sugata and Ishii [189], and Cranstoun and Anderson [190]. The "molecular clusters" discussed by Schubert et al. [191] may also represent islands. In any case, islands are considered to be characteristic of the transition period for low-temperature oxidation.

5.4.4. Driving Force for Island Growth

The formation of multiple oxide layers by place exchange or some other mode of ion movement implies the existence of a driving force for reaction beyond a monolayer. Ions must move through the initial chemisorbed layer, since oxygen no longer directly impinges upon the metal. The temperature is too low for thermally activated ion movement, so it is appropriate to invoke some other force. Fehlner and Mott [2] postulated the image force of an isolated oxygen ion adsorbed on the oxide monolayer as the driving force for reaction beyond a monolayer. Ritchie [192] has pointed out that stress and heat of chemisorption can also be involved. The image force simultaneously pulls the anion toward the metal and the cation toward the oxide–gas interface, thereby helping to overcome the activation energy for place exchange. The process accounts for both continued oxide growth and creation of cation sites at which further oxygen can be adsorbed. This model could be improved by consideration of electronic screening effects and lateral interactions in the adsorbed layer of oxygen.

The occurrence of place exchange requires that the bonds of the cation and anion in their respective positions be broken. The magnitude of the bond energies dominates the activation energy for place exchange, that is, formation of the reaction intermediate. The largest bond energy will be most significant and control the rate of the process.

Various metal–oxide and metal–metal energies have been collected in Table 5.2. The metal–oxide values have been taken from the compilations of Fehlner and Mott [2] and Kingery et al. [193] and plotted in Figure 3.3. This shows that the single oxide bond strengths are a periodic function of the cation atomic number.

The metal–oxide bond energies are based on the work of Sun and Huggins [194]. The determination of bond strength was carried out as follows. The heat of dissociation of an oxide into its gaseous atoms was calculated from heat of formation data. The heat of dissociation was then divided by the oxygen coordination number of the metal ion in the oxide. This number was obtained from X-ray data on crystals and is applicable to both the crystalline and vitreous states because approximately the same short-range order exists in both.

The metal–oxide bond is partially covalent in network formers and short range in character [195]. It is mainly ionic in network modifiers. One can, therefore, speak of single bond strengths dependent mainly on the nearest-neighbor environment.

On the other hand, metal–metal bonds are distributed in nature and have, therefore, no definite directionality. The long-range nature of the metallic bond means that next-nearest as well as nearest neighbors should be considered in calculating bond strengths. Tick and Witt [196] have discussed this point as it relates to metal surfaces.

We cannot then speak of a single metal–metal bond in the same sense that we speak of a single metal–oxide bond. Instead, the bonding of a metal atom to all of its neighbors must be considered when a metal ion dissolves in an oxide. The sublimation energy of a metal, defined as the energy required to remove atoms from the solid to the vapor phase, is a relative measure of this quantity. Sublimation, like metal dissolution in an oxide, occurs from the surface of a metal, preferably from a step or kink site where the work required to evaporate an atom can be decreased by dislocation strain energy or the presence of impurities. Of course, surface atoms have fewer nearest neighbors than bulk atoms [197], but this factor is taken into account during the measurement of heat of sublimation. Values included in Table 5.2 are considered to be the maximum possible values of metal–metal bond energy.

The ionization energy ϕ has not been included in this discussion because it applies to the separate process of electron transfer. It can be considered independently of ion movement as discussed in Chapter 2.

Place exchange during the transition period occurs between a partially incorporated anion and a cation bound to the surface of the metal, preferably at a step or kink site as pointed out by Mott [198]. Roberts [199] has discussed the evidence for the adsorption of oxygen ions on single cation sites at an oxide surface. His arguments support the assumption that only a single metal–oxide bond must be broken before place exchange can occur. On the other hand, the metal atom must be freed from the influence of all its surrounding metal atoms.

It is apparent that the largest bond strength will dominate the activation energy for place exchange. The thickness of oxide grown during the transition period will be determined by either the metal–metal or metal–oxide bond. However, the driving force for continued place exchange is still undefined. If we take the required activation energy as equal to the larger bond energy, M–O or M–M, then thermal energy at room temperature appears insufficient to explain the rate. Fehlner and Mott [2] have put forward a model, unproven as yet, in which the place exchange process will proceed with zero activation energy as long as the energy due to the image force (~2 eV or 46 kcal mole^{-1} for an ion one lattice spacing from

TABLE 5.2. COMPARATIVE METAL AND OXIDE BOND STRENGTHS

M in MO_x	(valence)	Heat of sublimation[a] at 25°C [kcal g-atom^{-1}]	Single oxide[b] bond strength [kcal mole^{-1}]	(coordination no.)
		Glass Formers		
B	(3)	—	119, 89	(3,4)
Si	(4)	84	106	(4)
Ge	(4)	—	108, 72	(4)
Ta	(5)	187	104	(6)
W	(6)	202	104	(6)
Nb	(5)	173	100	(6)
Al	(3)	77	101–79	(4)
Mo	(6)	159	92	(6)
P	(5)	—	111–88	(4)
V	(5)	123	112–90	(4)
As	(5)	—	87–70	(4)
Sb	(5)	63	85–68	(4)
Zr	(4)	146	81	(6)
		Intermediates		
Ti	(4)	113	73	(6)
Zn	(2)	31	72	(2)
Pb	(2)	47	73	(2)
Al	(3)	77	67–53	(6)
Sb	(3)	63	55	(6)
Th	(4)	138	64	(8)
Be	(2)	78	63	(4)
Zr	(4)	146	61	(8)
Cd	(2)	27	60	(2)
Sc	(3)	—	60	(6)
La	(3)	100	58	(7)

TABLE 5.2. (*Continued*)

M in MO_x (valence)		Heat of sublimation[a] at 25°C [kcal g-atom^{-1}]	Single oxide[b] bond strength [kcal mole^{-1}] (coordination no.)	
		Modifiers		
Y	(3)	102	50	(8)
Pt	(2)	135	50	(6)
Cr	(3)	95	47	(6)
Sn	(4)	72	46	(6)
Ga	(3)	65	45	(6)
Cu	(2)	81	44	(6)
In	(3)	58	43	(6)
Th	(4)	138	43	(12)
Rh	(3)	133	40	(6)
Pd	(2)	84	40	(6)
Pb	(4,2)	47	39,36	(6,4)
Mg	(2)	35	37	(6)
Li	(1)	39	36	(4)
Zn	(2)	31	36	(4)
Ba	(2)	—	33	(8)
Ca	(2)	42	32	(8)
Sr	(2)	39	32	(8)
Mn	(2)	67	33	(6)
Fe	(2)	100	32	(6)
Co	(2)	102	29	(6)
Ni	(2)	103	28	(6)
Cd	(2)	63	30,20	(4,6)
Na	(1)	26	20	(6)
K	(1)	21	13	(9)
Rb	(1)	20	12	(10)
Hg	(2)	15	11	(6)
Cs	(1)	19	10	(12)

[a] *Selected Values of Thermodynamic Properties of Metals and Alloys*, R. Hultgren, R. L. Orr, P. D. Anderson, and K. K. Kelley, John Wiley & Sons, New York, 1963.

[b] K-H. Sun, *J. Amer. Cer. Soc.* **30** (1947), 277. F. P. Fehlner and N. F. Mott, *Oxid. Metals* **2** (1970), 59.

the metal) is greater than the value of the critical bond energy which must be broken. The activation energy W is modified by the image force term yielding

$$W - \frac{2q^2a}{\kappa y^2} \tag{10}$$

as the the new activation energy, where y is the distance between the charges, q is the magnitude of charge, κ is the dielectric constant, and $2a$ is the ion jump distance. As y increases, Equation 10 becomes positive and place exchange must be thermally activated for continued oxide growth.

The work of Chou et al. [200] on the initial stages of oxide growth on lead agrees with a place exchange mechanism. Both measured and calculated thicknesses based on Equation 10 for the transition from place exchange to the beginning of logarithmic growth are in agreement.

To summarize then for the transition period, one or more layers of oxide grow and a charge builds up on the oxide surface. This charge can be observed as a change in the surface potential of the metal–oxide system. As ions form on the oxide, they are continually reacted in a process such as place exchange, but at a continually slower rate as the oxide thickens. During this time, a transition occurs at low temperatures from linear kinetics to logarithmic kinetics.

REFERENCES

1. N. Cabrera and N. F. Mott, *Rep. Prog. Phys.* **12** (1948–49), 163.
2. F. P. Fehlner and N. F. Mott, *Oxid. Metals* **2** (1970), 59.
3. N. D. S. Canning, D. Outka, and R. J. Madix, *Surface Sci.* **141** (1984), 240.
4. G. Ehrlich, in *Surface Phenomena of Metals*, S. C. I. Monograph No. 28, Society of Chemical Industry, London, 1968, p. 13.
5. *Surface Contamination* (Proc. Symp. Surface Contamination, Genesis, Detection, and Control, Washington, 1978), K. L. Mittal, Ed., Plenum Press, New York, 1979.
6. C. Leung and R. Gomer, *Surface Sci.* **59** (1976), 638.
7. M. W. Roberts and B. R. Wells, *ibid.* **8** (1967), 453; **15** (1969), 325.
8. M. W. Roberts and C. S. McKee, *Chemistry of the Metal-Gas Interface*, Clarendon Press, Oxford, 1978.
9. A. U. MacRae and L. H. Germer, *Ann. N.Y. Acad. Sci.* **101** (1963), 627.
10. A. U. MacRae and L. H. Germer, *Phys. Rev. Lett.* **8** (1962), 489.
11. A. U. MacRae, *Surface Sci.* **2** (1964), 522.
12. H. E. Farnsworth, *Ann. N.Y. Acad. Sci.* **101** (1963), 658.
13. J. J. Lander and J. Morrison, *ibid.*, p. 605.
14. J. J. Lander, *Surface Sci.* **1** (1964), 125.
15. J. E. Inglesfield and B. W. Holland, in *The Chemical Physics of Solid Surfaces and Heterogeneous Catalysis*, Vol. 1, D. A. King and D. P. Woodruff, Eds., Elsevier, New York, 1981, p. 310.

16. *Interactions on Metal Surfaces*, R. Gomer, Ed., Springer-Verlag, New York, 1975.
17. *Adsorption-Desorption Phenomena*, Proc. 2nd Intern. Conf. Florence-1971, F. Ricca, Ed., Academic Press, New York, 1972.
18. G. A. Somorjai, *Principles of Surface Chemistry*, Prentice-Hall, Englewood Cliffs, NJ, 1972.
19. D. O. Hayward and B. M. W. Trapnell, *Chemisorption*, Butterworths, London, 1964.
20. D. M. Young and A. D. Crowell, *Physical Adsorption of Gases*, Butterworths, London, 1962.
21. G. Ehrlich, in *Metal Surfaces: Structure, Energetics, and Kinetics* (Papers Joint Seminar Amer. Soc. Metals and Met. Soc. AIME, 1962), American Society for Metals, Metals Park, OH, 1963, p. 221.
22. J. Lepage and D. Paulmier, *Surface Sci.* **104** (1981), 569.
23. N. Cabrera, in *Semiconductor Surface Physics* (Proc. Conf. Physics Semicond. Surf., Philadelphia, 1956), R. H. Kingston, Ed., University of Pennsylvania Press, Philadelphia, 1957, p. 327.
24. M. L. Hair, *Infrared Spectroscopy in Surface Chemistry*, Marcel Dekker, New York, 1967.
25. A. T. Fromhold, Jr. *Theory of Metal Oxidation, Vol. 1—Fundamentals*, North-Holland, New York, 1976. (a) p. 7; (b) p. 86.
26. C. Benndorf, B. Egert, C. Nöbl, H. Seidel, and F. Thieme, *J. Vac. Sci. Technol.* **17** (1980), 115.
27. G. A. Somorjai, *Chemistry in Two-Dimensions: Surfaces*, Cornell University Press, Ithaca, NY, 1981. (a) p. 37ff; (b) p. 145ff; (c) p. 283ff.
28. T. Wolfram, *J. Vac. Sci. Technol.* **18** (1981), 428.
29. J. Eickmans, A. Goldmann, and A. Otto, *Surface Sci.* **127** (1983), 153.
30. G. Ehrlich, *Ann. N.Y. Acad. Sci.* **101** (1963), 221, 722.
31. T. Oguri, *J. Phys. Soc. Japan* **19** (1964), 83.
32. T. W. Hickmott, *J. Chem. Phys.* **32** (1960), 810.
33. T. Oguri, *J. Phys. Soc. Japan* **19** (1964), 77.
34. P. A. Redhead, *Trans. Faraday Soc.* **57** (1961), 641.
35. G. Ehrlich, *J. Chem. Phys.* **34** (1961), 39.
36. J. A. Becker, *Solid State Phys.* **7** (1958), 379.
37. T. E. Madey, J. J. Czyzewski, and J. T. Yates, Jr., *Surface Sci.* **49** (1975), 465.
38. J. L. Gland, *ibid.* **93** (1980), 487.
39. J. L. Gland, B. A. Sexton, and G. B. Fisher, *ibid.* **95** (1980), 587.
40. A. Spitzer and H. Lüth, *ibid.* **118** (1982), 121, 136.
41. T. G. Mills and F. A. Kroger, *J. Electrochem. Soc.* **120** (1973), 1582.
42. J. H. Lunsford, *Catal. Rev.* **8** (1973), 135.
43. E. G. Derouane and V. Indovina, *Chem. Phys. Lett.* **14** (1972), 455.
44. R. B. Clarkson and A. C. Cirillo, Jr., *J. Vac. Sci. Technol.* **9** (1972), 1073.
45. R. D. Iyengar and M. Codell, *Advanc. Colloid Interface Sci.* **3** (1972), 365.
46. M. Formenti, H. Courbon, F. Juillet, A. Lissatchenko, J. R. Martin, P. Meriaudeau, and S. J. Teichner, *J. Vac. Sci. Technol.* **9** (1972), 947.
47. T. Smith, *Surface Sci.* **38** (1973), 292.
48. A. Benninghoven et al., *ibid.* **39** (1973), 397, 416, 427.
49. A. Benninghoven, *ibid.* **28** (1971), 541; **35** (1973), 427.
50. R. W. Joyner and M. W. Roberts, *Chem. Phys. Lett.* **28** (1974), 246.
51. See reviews in *Surface Sci.* **25** (1971), 1–223.
52. *Proc. 7th Intern. Vac. Congr. and 3rd Intern. Conf. Solid Surfaces*, R. Dobrozemsky et al., Eds., Vienna, 1977.

53. *Methods of Surface Analysis*, A. W. Czanderna, Ed., Elsevier, New York, 1975.
54. *Applied Surface Analysis*, T. L. Barr and L. E. Davis, Eds., Amer. Soc. for Testing and Materials, Philadelphia, 1980, STP No. 699.
55. *Electron Spectroscopy for Surface Mapping*, H. Ibach et al., Eds., Springer-Verlag, New York, 1977.
56. F. Abelès, *Thin Solid Films* **34** (1976), 291.
57. C. R. Brundle, *J. Electron Spectrosc. Relat. Phenom.* **5** (1974), 291.
58. H. J. Mathieu, M. Datta, and D. Landolt, *J. Vac. Sci. Technol.* **A3** (1985), 331.
59. E. W. Müller, in *Advances in Electronics and Electron Physics*, Vol. 13, L. Marton, and C. Marton, Eds., Academic Press, New York, 1960, p. 83.
60. D. P. Woodruff, in *The Chemical Physics of Solid Surfaces and Heterogeneous Catalysis*, Vol. 1, D. A. King and D. P. Woodruff, Eds., Elsevier, New York, 1981, p. 81.
61. E. A. Wood, *J. Appl. Phys.* **35** (1964), 1306.
62. P. Mark, in *Surfaces and Interfaces I*, J. J. Burke et al., Eds., Syracuse University Press, Syracuse, NY, 1967, p. 103.
63. G. Ertl and J. Kueppers, *Low Energy Electrons and Surface Chemistry*, Verlag Chemie, Weinheim, 1974.
64. J. P. Biberian and G. A. Somorjai, *J. Vac. Sci. Technol.* **16** (1979), 2073.
65. G. L. Hunt and I. M. Ritchie, *J. Chem. Soc. Faraday Trans. I* **68** (1972), 1413.
66. G. A. Hope and I. M. Ritchie, *Thin Solid Films* **34** (1976), 111.
67. F. P. Fehlner, *J. Electrochem. Soc.* **115** (1968), 726.
68. F. P. Fehlner, *Nature* **210** (1966), 1035.
69. F. P. Fehlner, *Trans. 3rd Intern. Vac. Cong.*, Vol. 2, Pergamon Press, Oxford, 1966, p. 691.
70. F. P. Fehlner, *J. Appl. Phys.* **38** (1967), 2223.
71. J. G. Swanson, D. S. Campbell, and J. C. Anderson, *Thin Solid Films* **1** (1967/68), 325.
72. C. A. Neugebauer and M. B. Webb, *J. Appl. Phys.* **33** (1962), 74.
73. W-K. Chu, J. W. Mayer, and M. A. Nicolet, *Back-Scattering Spectrometry*, Academic Press, New York, 1978.
74. A. M. Horgan and D. A. King, *Surface Sci.* **23** (1970), 259.
75. E. Fromm, H. Duppel, and U. Bauder, *Thin Solid Films* **33** (1976), 323.
76. D. E. Fowler and J. M. Blakely, *J. Vac. Sci. Technol.* **20** (1982), 930.
77. A. Krozer and B. Kasemo, *Surface Sci.* **97** (1980), L339.
78. I. P. Batra and L. Kleinman, *J. Electron Spectrosc. Relat. Phenom.* **33** (1984), 175.
79. C. A. Carosella and J. Comas, *Surface Sci.* **15** (1969), 303.
80. *Proc. 7th Intern. Vac. Cong. and 3rd Intern. Conf. Solid Surfaces*, R. Dobrozemsky et al., Eds., Vienna, 1977; E. Fromm, p. 889; W. H. Weinberg et al., p. 1151; F. H. P. M. Habraken et al., p. 877; M. A. Chesters and J. C. Riviere, p. 873.
81. C. Wang and R. Gomer, *ibid.*, p. 1155.
82. D. A. King, *ibid.*, p. 769.
83. H. P. Bonzel, *ibid.*, p. 691; K. Besocke and S. Berger, *ibid.*, p. 893; G. Pirug et al., *ibid.*, p. 907.
84. E. Bauer and H. Poppa, *Surface Sci.* **127** (1983), 243.
85. A. F. Armitage and D. P. Woodruff, *ibid.* **114** (1982), 414.
86. S. Tougaard, P. Morgen, and J. Onsgaard, *ibid.* **111** (1981), 545.
87. J. C. Dash, *Scientific American* **228** No. 5 (1973), 30.
88. D. M. Collins and W. E. Spicer, *Surface Sci.* **69** (1977), 85.
89. J. C. Phillips, *Physics Today* **23** (1970), 23.
90. G. Ehrlich, in *1961 Trans. 8th Vacuum Symp. and 2nd Intern. Cong.*, Pergamon Press, New York, 1962, p. 126.

91. G. Ehrlich, in *Interatomic Potentials and Simulation of Lattice Defects* (Battelle Inst. Mater. Sci. Colloq., 6th, Seattle, 1971), P. C. Gehlen et al., Eds., Plenum Press, New York, 1972, p. 573.
92. W. Rühl, *Z. Physik* **186** (1965), 190.
93. H. E. Farnsworth and H. H. Madden, Jr., *J. Appl. Phys.* **32** (1961), 1933.
94. A. Redondo, W. A. Goddard, III, C. A. Swarts, and T. C. McGill, *J. Vac. Sci. Technol.* **19** (1981), 498.
95. L. Surnev and M. Tikhov, *Surface Sci.* **123** (1982), 505.
96. E. Fromm and O. Mayer, *ibid.* **74** (1978), 259.
97. P. H. Holloway and J. B. Hudson, *ibid.* **43** (1974), 123, 141.
98. A. J. Pignocco and G. E. Pellissier, *ibid.* **7** (1967), 261.
99. Th. Heumann and I-H. Moon, *ibid.* **24** (1971), 370.
100. W. N. Unertl and J. M. Blakely, *ibid.* **69** (1977), 23.
101. D. E. Fowler and J. M. Blakely, *J. Vac. Sci. Technol.* **20** (1982), 930.
102. S. Ekelund and C. Leygraf, *Surface Sci.* **40** (1973), 179.
103. C. Y. Su, I. Lindau, P. R. Skeath, I. Hino, and W. E. Spicer, *ibid.* **118** (1982), 257.
104. J. W. May and L. H. Germer, *ibid.* **11** (1968), 443.
105. H. D. Hagstrum and G. E. Becker, *J. Chem. Phys.* **54** (1971), 1015.
106. T. E. Madey and J. T. Yates, Jr., *J. Vac. Sci. Technol.* **8** (1971), 525.
107. A. E. Lee and B. A. Pethica, *Proc. Roy. Soc. (London)* **A309** (1969), 141.
108. U. R. Evans, *Metallic Corrosion Passivity and Protection*, 2nd ed., Edward Arnold, London, 1946.
109. W. H. Krueger and S. P. Pollack, *Surface Sci.* **30** (1972), 263.
110. A. F. Carley and M. W. Roberts, *Proc. Roy. Soc. (London)* **A363** (1978), 403.
111. S. Dushman, *Scientific Foundations of Vacuum Technique*, 2nd ed., John Wiley & Sons, New York, 1962, p. 389ff.
112. R. Sau and J. B. Hudson, *Surface Sci.* **102** (1981), 239.
113. P. B. Sewell, D. F. Mitchell, and M. Cohen, *ibid.* **33** (1972), 535.
114. H. Windawi, *J. Vac. Sci. Technol.* **18** (1981), 660.
115. S. Mróz, C. Koziol, and J. Kolaczkiewicz, *Vacuum* **26** (1976), 61.
116. R. Klein, R. Siegel, and N. E. Erickson, *J. Vac. Sci. Technol.* **16** (1979), 489.
117. G. A. Haas and H. F. Gray, *J. Appl. Phys.* **46** (1975), 3885.
118. A. U. MacRae, *Science* **139** (1963), 379; *Surface Sci.* **1** (1964), 319.
119. L. H. Germer, R. H. Stern, and A. U. MacRae, in *Metal Surfaces: Structure, Energetics and Kinetics* (Papers Joint Seminar Amer. Soc. Metals and Met. Soc. AIME, 1962) American Society for Metals, Metals Park, OH, 1963, p. 287.
120. K. R. Lawless, *Rep. Prog. Phys.* **37** (1974), 231.
121. C. T. Kirk, Jr. and E. E. Huber, Jr., *Surface Sci.* **9** (1968), 217; **5** (1966), 447.
122. H. L. Yu, M. C. Muñoz, and F. Soria, *ibid.* **94** (1980), L184.
123. D. F. Mitchell, R. B. Sewell, and M. Cohen, *ibid.* **69** (1977), 310; **61** (1976), 355.
124. P. H. Holloway, *J. Vac. Sci. Technol.* **18** (1981), 653.
125. A. Benninghoven et al., in *Proc. 7th Intern. Vac. Cong. and 3rd Intern. Conf. Solid. Surfaces*, R. Dobrozemsky, et al., Eds., Vienna, 1977, pp. 723, 2577.
126. C. Wagner, *Corrosion Sci.* **10** (1970), 641.
127. M. W. Roberts, *Trans. Faraday Soc.* **57** (1961), 99.
128. W. H. Weinberg, S. P. Withrow, and W. F. Egelhoff, Jr., *J. Vac. Sci. Technol.* **14** (1977), 482.
129. W. Heiland and E. Taglauer, *ibid.* **9** (1972), 620.
130. S. Evans, J. Pielaszek, and J. M. Thomas, *Surface Sci.* **55** (1976), 644.
131. S. Evans et al., *J. Chem. Soc. Faraday Trans. II* **71** (1975), 313, 1044; *Faraday Disc. Chem. Soc.* **58** (1975), 97; **60** (1976), 102.

132. W. G. Dorfeld, J. B. Hudson, and R. Zuhr, *Surface Sci.* **57** (1976), 460.
133. H. Heyne and F. C. Tompkins, in *Surface Phenomena of Metals*, S.C.I. Monograph No. 28, Society of Chemical Industry, London, 1968, p. 339.
134. C. Nyberg, *Surface Sci.* **65** (1977), 389.
135. Y. Sakisaka, H. Kato, and M. Onchi, *ibid.* **120** (1982), 150.
136. Z. Hurych, R. L. Benbow, and J. C. Shaffer, *J. Vac. Sci. Technol.* **14** (1977), 391.
137. L. H. Dubois, *Surface Sci.* **119** (1982), 399.
138. J. B. Bignolas and M. Bujor, *ibid.* **108** (1981), L453.
139. S. A. Flodstrom, R. Z. Bachrach, R. S. Bauer, and S. B. M. Hagström, *Phys. Rev. Lett.* **37** (1976), 1282.
140. M. Bujor, L. A. Larson, and H. Poppa, *J. Vac. Sci. Technol.* **20** (1982), 392.
141. C. Y. Su, P. R. Skeath, I. Lindau, and W. E. Spicer, *Surface Sci.* **118** (1982), 248.
142. F. Watari and J. M. Cowley, *ibid.* **105** (1981), 240.
143. C. R. Brundle, T. J. Chuang, and K. Wandelt, *ibid.* **68** (1977), 459.
144. R. B. Moyes and M. W. Roberts, *J. Catalysis* **49** (1977), 216.
145. J. W. Rogers, Jr. and M. L. Knotek, *J. Vac. Sci. Technol.* **20** (1982), 414.
146. G. Hollinger and F. J. Himpsel, *ibid.* **A1** (1983), 640.
147. N. Lieske and R. Hezel, *Thin Solid Films* **61** (1979), 197.
148. S. Sato, T. Inoue, and H. Sasaki, *ibid.* **86** (1981), 21.
149. P. C. Karulkar and J. E. Nordman, *J. Vac. Sci. Technol.* **17** (1980), 462.
150. D. T. Larson, *ibid.* **17** (1980), 55.
151. D. T. Ling, J. N. Miller, P. Pianetta, D. L. Weissman, I. Lindau, and W. E. Spicer, *ibid.* **21** (1982), 47.
152. B. D. Padalia, J. K. Gimzewski, and S. Affrossman, *Surface Sci.* **61** (1976), 468.
153. J. M. Baker and J. L. McNatt, *J. Vac. Sci. Technol.* **9** (1972), 792.
154. W. Färber and P. Braun, *Surface Sci.* **41** (1974), 195.
155. G. Hecht, J. Herberger, and C. Weissmantel, *J. Vac. Sci. Technol.* **7** (1970), 547.
156. C. Weissmantel, G. Hecht, J. Herberger, R. Glauche, and I. Wagner, *Krist. Tech.* **5** (1970), 299.
157. J. Lapujoulade, Y. Le Cruër, M. Lefort, Y. Lejay, and E. Maurel, *Surface Sci.* **118** (1982), 103.
158. V. D. Vankar, R. W. Vook, and B. C. DeCooman, *Thin Solid Films* **102** (1983), 313.
159. A. T. Fromhold, Jr., *J. Electrochem. Soc.* **127** (1980), 411.
160. A. E. Lee, *Surface Sci.* **47** (1975), 191.
161. R. D. Young and D. C. Schubert, *J. Chem. Phys.* **42** (1965), 3943.
162. J. Bardolle and J. Bénard, *Compt. rend.* **232** (1951), 231; **239** (1954), 706.
163. T-M. Lu, G-C. Wang, and M. G. Lagally, *Surface Sci.* **92** (1980), 133.
164. J. Bénard, in *Oxidation of Metals and Alloys* (ASM Seminar, Cleveland, October, 1970), D. L. Douglass, Ed., American Society for Metals, Metals Park, OH, 1971, p. 1.
165. J. Bénard, F. Grønlund, J. Oudar, and M. Duret, *Z. Elektrochem.* **63** (1959), 799, 804.
166. F. Grønlund, *J. Chim. Phys.* **53** (1956), 660.
167. J. Bardolle and J. Bénard, *Rev. Met.* **49** (1952), 613.
168. J. Bénard, *Acta Met.* **8** (1960), 272.
169. A. T. Gwathmey and K. R. Lawless, in *The Surface Chemistry of Metals and Semiconductors*, H. C. Gatos et al., Ed., John Wiley & Sons, New York, 1960, p. 483.
170. H. Fischmeister, in *Reactivity of Solids* (Proc. 4th Intern. Symp. Reactivity Solids, Amsterdam, 1960) Elsevier, Amsterdam, 1961, p. 195.
171. J. L. Domange, *J. Vac. Sci. Technol.* **9** (1972), 682.
172. J. M. Ritchie, *J. Australian Inst. Metals* **11** (1966), 242.
173. J. Oudar, *J. Vac. Sci. Technol.* **9** (1972), 657.

174. G. E. Rhead, *ibid.* **13** (1976), 603.
175. S. Wagener, *Brit. J. Appl. Phys.* **1** (1950), 225; **2** (1951), 132.
176. T. Arizumi and S. Kotani, *J. Phys. Soc. Japan* **7** (1952), 300.
177. P. della Porta, *Vacuum* **4** (1954), 464.
178. R. N. Bloomer, *Brit. J. Appl. Phys.* **8** (1957), 40, 321.
179. M. S. Cohen, *Acta Met.* **8** (1960), 356.
180. R. R. Addiss, Jr., *ibid.* **11** (1963), 129.
181. A. G. Revesz and F. P. Fehlner, *Oxid. Metals* **15** (1981), 297.
182. W. H. Orr, Thesis Cornell University, No. 62-5965, University Microfilms Inc., Ann Arbor, MI, 1962.
183. A. J. Melmed and J. J. Carroll, *J. Vac. Sci. Technol.* **10** (1973), 164.
184. G. K. Hall and C. H. B. Mee, *Surface Sci.* **28** (1971), 598.
185. C. S. McKee, L. V. Renny, and M. W. Roberts, *ibid.* **75** (1978), 92.
186. J. M. Thomas, *Advan. Catal. Relat. Subj.* **19** (1969), 293.
187. P. B. Barna, *Thin Solid Films* **85** (1981), 263.
188. S. S. Brenner and W. J. McVeagh, *J. Electrochem. Soc.* **115** (1968), 1247.
189. E. Sugata and S. Ishii, *Surface Sci.* **24** (1971), 663.
190. G. K. L. Cranstoun and J. S. Anderson, *Nature* **219** (1968) 365; *Surface Sci.* **32** (1972), 397.
191. C. C. Schubert, C. L. Page, and B. Ralph, *Electrochim. Acta* **18** (1973), 33.
192. I. M. Ritchie, in *Chemisorption and Reactions on Metallic Films*, Vol. 2, J. R. Anderson, Ed., Academic Press, London, 1971, p. 257.
193. W. D. Kingery, H. K. Bowen, and D. R. Uhlmann, *Introduction to Ceramics*, 2nd ed., John Wiley & Sons, New York, 1976, p. 99.
194. K-H. Sun and M. L. Huggins, *J. Phys. Colloid Chem.* **51** (1947), 438.
195. L. Pauling, *The Nature of the Chemical Bond*, 2nd ed., Cornell University Press, Ithaca, NY, 1948.
196. P. A. Tick and A. F. Witt, *Surface Sci.* **26** (1971), 165.
197. E. U. Condon, *Amer. J. Phys.* **22** (1954), 224.
198. N. F. Mott, *Trans. Faraday Soc.* **36** (1940), 472; **43** (1947), 429.
199. M. Wyn Roberts, *Quart. Rev.* (*London*) **16** (1962), 71.
200. N. J. Chou, J. M. Eldridge, R. Hammer and D. Dong, *J. Electronic Mater.* **2** (1973), 115.

6

THE ROLE OF INTERFACES

Gas-phase oxidation begins at the interface between a metal or semiconductor and an oxidant, for example, oxygen. It continues with the establishment of a second interface between the product oxide and oxygen. Incorporation of cations in the oxide occurs at the metal–oxide interface, while incorporation of anions occurs at the oxide–gas interface. Thus, crystallographic structure, phase change in the metal, Curie temperature [1], and oxygen pressure can affect the rate of oxidation.

6.1. NATURE OF THE METAL–OXIDE INTERFACE

The very creation of a surface, that is, a discontinuity in the solid, gives rise to electronic states at the surface. The broken or dangling bonds of an unreconstructed surface are called Tamm states. They contain unbonded electrons. In a metal, the surface atoms may relax and sometimes reconstruct before chemisorption so that the nondirectional metallic bonding can satisfy the discontinuity. In covalently bonded semiconductors, surface–atom rearrangement is more difficult and the surface bonds are satisfied by both reconstruction and chemisorption. As a result, when such a surface is oxidized, a great many of the dangling bonds are saturated. The number remaining unsaturated depends on the lattice mismatch between a crystalline oxide and the underlying substrate or on the flexibility of bonding between a vitreous oxide and the substrate. It has been suggested that the dangling bonds are related to the "fast" surface states, now called interface-trapped charge, discussed in silicon technology [2]. See Chapter 9.

Incorporation of excess cations or impurities in the oxide leads to unsaturated bonds. Many oxides in equilibrium with a metal appear to have a distribution of excess cations extending from the metal–oxide interface and exponentially decreasing to the intrinsic portion of the oxide. On the other hand, silicon–silicon dioxide can have an abrupt interface. This is discussed below.

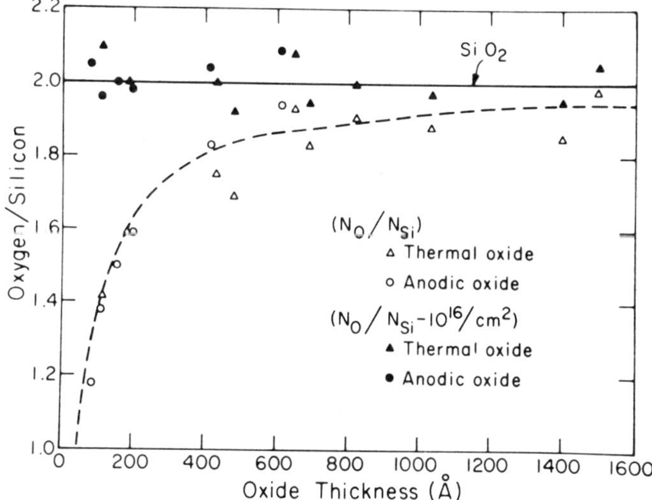

Figure 6.1. Ratio of oxygen to silicon versus oxide thickness for thermal and anodic oxides grown on silicon. After Sigmon et al. [17]. Published by permission of the copyright holder, the American Institute of Physics.

Heine and Pryor [3,4] have extensively studied the stoichiometry of anodic oxide films on aluminum. Electrical measurements in solution have led them to conclude that a cation excess exists at the metal–oxide interface. In addition, the effects of anion incorporation at the oxide–electrolyte interface have been examined [5,6,7,8,9].

Further evidence on this point is found in work on tantalum oxide films [10]. Ellipsometer measurements of tantalum oxide grown anodically [8] or by plasma oxidation [11] can be interpreted in terms of a double-layer film. The inner film is assumed to have a higher refractive index, as would be the case for a metal-rich layer. However, there is some dispute over this interpretation [12] especially when the anolyte is changed [13].

Two-layer films on anodized cobalt are reported by Ohtsuka and Sato [14]. A model for such film growth has been presented by Fromhold [15] and the growth kinetics calculated by Fromhold and Sato [16].

Similar interfacial effects are found in other oxides. The nonstoichiometry of SiO_2 formed by oxidation and anodization is shown in Figure 6.1 as determined by Sigmon et al. [17]. Flitsch and Raider [18] using photoelectron spectroscopy detected a similar interfacial nonstoichiometry. With time, improvements in oxidation methods and analytical techniques showed that the interfacial region could be thin [19,20,21,22] for SiO_2. The interface is considered to be abrupt, that is, 10 Å or less

[23,24,25,26,27]. Bauer and Bachrach [28] find that the intermediate state SiO is found at the Si/SiO_2 interface formed at 700°C, but not at one formed at 25°C. See Chapter 9 for further discussion of this point.

Impurities play a vital role in the creation of both "fast" and "slow" surface states. For instance, hydrogen can react with a dangling bond at a silicon–oxide interface to form a hydride [29], thus destroying a "fast" surface state. Cation impurities such as sodium near the same interface can conceivably act as "slow" surface states. These in turn can be compensated by ions of opposite charge such as chloride.

Surface states exist in practice for semiconductors, and in theory they could exist for metals. It turns out to be far easier to study them on semiconductors such as silicon and germanium. This is so because the small number of free carriers in the semiconductor strongly reflects the effect of the small density of surface states. Thus, defects can be studied electronically, for example, by capacitance–voltage measurements [2] or lifetime studies [30]. Silicon oxide is a particularly interesting material because it is the best network-forming oxide known. The flexibility of its structure allows the oxide to bond to most of the silicon surface atoms, leading to a minimum number of surface states [31].

6.1.1. Exoelectron Emission

Exoelectron emission [32,33] at room temperature is a process in which electrons are emitted from a metal surface in vacuum. Several explanations of the process have been advanced, one of which relates to defects in the oxide [34]. The presence of an oxidant often appears to be necessary for the effect. Both photostimulation and abrasion increase the magnitude of the electron current.

Several different types of exoelectron emission have been noted. For oxides thicker than 450 Å, emission has been ascribed to cracking of the oxide [35]. Oxide regrowth on the fresh metal exposed by the crack as well as charge separation between the opposite sides of the crack may be involved in the emission process. For thinner oxides, slip planes in the metal may penetrate the oxide, providing a path for electron escape. Phase transformations in amorphous solids have been detected based on the temperature dependence of photostimulated exoelectron emission from Se and B_2O_3 [36].

Exoelectron emission has been noted directly during the chemisorption of oxygen on a clean metal surface [34,37]. For silicon, no thermally stimulated exoelectron emission was found until the surface was exposed to oxygen and covered with SiO_2 [38]. The actual details of this process are still under investigation. However, a correlation of exoelectron emission with work function [39], oxide strain, "glow" curves, and thermal luminescence of the oxide has been shown [34]. A maximum in emission occurs at a minimum in work function for magnesium [39]. A distinction between exoelectrons emitted from Mg in the dark at low oxygen exposures and photostimulated emission is made by Allen et al. [40].

6.1.2. Possible Role of Surface States

A relationship between oxidation and surface states can be suggested tentatively as follows. If cation incorporation at the metal–oxide interface is rate controlling, then the most weakly bound cations will be those associated with dangling bonds, especially at kink sites and steps on the metal surface. These sites will correspond to fast surface states and may be related to the quantity N in Equation 21 of Chapter 2. The number of fast surface states should increase as the level of surface disorder (kink sites, interstitials, grain boundaries) increases. Thus, polycrystalline metals and semiconductors are expected to have a larger value of N than single crystals and metallic glasses.

On the other hand, when local ionic equilibrium is established at the metal–oxide interface for thicker oxides, slow surface states may exist in the form of excess dissolved cations with associated dangling bonds. As noted above, a cation excess has often been found at the oxide–metal interface. It may be speculated that, in this case, the value of N for cation movement is related to a virtual cation source existing in the oxide near the oxide–metal interface. On the other hand, if anions are the mobile species, then N should be related to the number of partially incorporated anions at the oxide–gas interface.

Studies of catalytic reactions show that surface states exist at the oxide–gas interface as well as at the metal–oxide interface for thick oxides [41,42]. Such states are referred to as active sites [43]. They have been studied by nuclear magnetic resonance [44], infrared [45] and photoelectron spectroscopy [46], and other methods [47].

6.2. NATURE OF THE OXIDE–GAS INTERFACE

Charged species present at the oxide–gas interface are discussed in Chapter 5. The behavior of these charges has been studied using surface potential, that is, work function, measurements. Generally a vibrating capacitor has been employed [48], although other techniques are available [49,50].

The work function (negative of the surface potential) of a metal can experience several changes during oxidation. See Figure 5.10. Two basic processes appear to dominate the magnitude and sign of the changes in the presence of dry oxygen: the formation of a chemisorbed layer of oxygen and incorporation of this oxygen into the metal surface [51]. Physisorption follows chemisorption, but affects surface potential to a lesser degree.

Most metals at low temperatures show an overall increase in work function ϕ_v (decrease in surface potential) upon adsorbing oxygen, indicating that a negatively charged surface layer has formed. Water adsorption can interfere with observing this increase since it leads to a decrease in work function as observed on Al [52], Ni [53], and Cu [54,55,56]. Bonding of water occurs mainly through the oxygen atom. Reaction results in hydroxyl formation, both on bare and oxygen-covered aluminum [57]. Oxygen plays a role in the hydroxylation of zinc [58] and copper [59] while hydroxyls can slow the reaction of oxygen with iron [60].

The increase in ϕ_v due to oxygen adsorption and electron capture, up to monolayer coverage as shown in Figure 5.10, has several distinct regions [61]. An initial small decrease in work function (not shown in Figure 5.10) has been ascribed to early incorporation of oxygen into the metal [62,63]. This is followed by an increase in work function to the point of stable oxide film formation. The rate of change in ϕ_v during this increase is directly proportional to oxygen pressure. The extent of work function change can be both temperature and pressure dependent. See Table 6.1 where maximum changes in surface potential, usually for approximately monolayer coverage, are tabulated.

Sometimes, a decrease in work function is noted midway in the above sequence. It has been ascribed to geometric field enhancement due to formation of discrete oxide islands [64] on iron. The decrease in work function is most readily observed on single crystal metals where the density of nuclei is small, for example, nickel [65,66], leading to a limited number of islands. Lateral growth of the islands can result in an initial oxide film which is thicker than a monolayer.

Other workers have ascribed the decrease in ϕ_v to reconstruction of, that is, incorporation of oxygen into, a Cu (100) surface [67]. A detailed examination of the phenomenon for Cr (110) has been carried out at 300°K in which work function measurements were cross correlated with electron energy loss spectroscopy [68]. Four stages of oxygen interaction were correlated with oxygen coverage: dissociative chemisorption to 2 L, incorporation of oxygen adatoms between 2 and 6 L, rapid oxidation from 6 to 15 L leading to Cr_2O_3 formation, and slow thickening of Cr_2O_3 above 15 L. Since the same process, taken to be place exchange in Chapter 5, may lead to both incorporation and island growth, similar explanations for decreasing work functions may be invoked for both cases.

Magnesium shows a similar sequence of chemisorption followed by oxygen incorporation, island growth, oxide formation, and oxide thickening [69,70].

TABLE 6.1. CHANGES IN SURFACE POTENTIAL FOR OXYGEN ADSORPTION ON METALS

Metal	Form	Reference	T (°C)	P_{O_2} (Torr)	ΔSP (eV)	Remarks
Ag	(100)	106	22	10^{-5} to 10^{-7}	+0.17	850-L exposure
		106	140	10^{-5} to 10^{-7}	−0.29	850 L
	(110)	106	25	10^{-5} to 10^{-7}	−0.85	At "saturation"; small function of T
	(111)	106	25	10^{-5}	−0.04	
	(111)	107	150	10^{-1}	−0.25	
Al	Film	71	−195	$\sim 10^{-4}$	−0.8	150 min
	Film	71	−183	$\sim 10^{-4}$	−0.6	150 min
	Film	52,102	25	10^{-3} to 10	−0.3 to −0.7	P dependent
	Film	73	23	$\sim 10^{-2}$	−0.1	$\Delta SP = -0.6$ at saturation
	Film	73	−195	$\sim 10^{-2}$	−0.83	Dose saturated
	Polyrod	108	26	10^{-8} to 10^{-6}	+0.20 to +0.07	P dependent—saturated
	Polyrod	51	26	2×10^{-8}	+0.2	150 min
	Polyrod	109	−30	10^{-6}	−0.1	50 min
	Polyrod	109	200	10^{-6} to 10^{-7}	+0.4	75 min
	Polyrod	51	250	10^{-7}	+0.48	20 min
	Polyrod	51	250	7	+0.40	20 min, ΔSP is dependent on preexposure conditions
	(100)	110	27	5×10^{-7}	+1.09	500 L
	(110)	110	27	5×10^{-7}	+0.49	500 L
	(111)	110	27	5×10^{-7}	+0.19	500 L
	(100)	111	25	10^{-7}	+1.3	325 L
	(110)	111	25	10^{-7}	+0.8	325 L
	(111)	111	25	10^{-7}	+0.2	325 L
	Polyrod	112	250	3×10^{-9}	+0.30	160 min
	Polyrod	112	250	1×10^{-7}	+0.49	38 min

TABLE 6.1. (*Continued*)

Metal	Form	Reference	T (°C)	P_{O_2} (Torr)	ΔSP (eV)	Remarks
Ba	Film	113	RT	Doses	+0.32	Saturated
Co	Film	62	17	2×10^{-9}	−0.01	1×10^{15} molecules cm^{-2}, "N" curve
Cr	Film	114	−195, 25	10^{-2}	−2.1	Saturated
	(110)	68	27	$1–5 \times 10^{-7}$	−0.32	15 L, "N" curve
	(111)	115	25	10^{-9} to 10^{-7}	−0.9	~ 9 Å oxide
Cu	Film	116	25	$\sim 10^{-6}$	−0.3	30 L; ΔSP = −0.14 at 10^5 L
	Film	114	−195	10^{-2}	−0.55	Saturated
	Film	114	25	10^{-2}	−0.63	Saturated
	Film	114	25	40	−0.85	Saturated
	(100)	67	27	10^{-8} to 10^{-5}	−0.33	300 L, *SP* increases at long exposure
	(100)	67	300	10^{-8} to 10^{-5}	−0.30	75 L, *SP* increases at long exposure
					−0.03	700 L
	(111)	117	25, −196		−0.13	4×10^{16} O_2 cm^{-2} exposure
	(100)	117	25, −196		−0.39	4×10^{16} O_2 cm^{-2} exposure
	(110)	117	25		−0.68	7×10^{17} O_2 cm^{-2} exposure
	(110)	117	−196		−0.36	4×10^{16} O_2 cm^{-2} exposure
	(100)	118	27		−0.27	1000 L
			−173		−0.43	200 L

Cu	(110)	118	27		−0.37	100 L
			−173		−0.27	20 L
	(111)	119	27		0.0	0 to 1000 L
			−173		+0.40	1000 L
Er	Film	120	27	Doses	−0.4	For $\theta \cong 0.5$; $\Delta SP = +0.17$ at $\theta = 1.3$
Fe	Film	114	−195	10^{-2}	−1.55	Saturated
	Film	62	17	2×10^{-9}	−0.2	10^{15} molecules cm^{-2}; "N" curve
	(011)	64	25	5×10^{-9} to 5×10^{-6}	−0.5	50 L
	Film	114	25	10^{-2} to 50	−1.5	Saturated
Ge	(100)	121	27	5×10^{-7} to 5×10^{-8}	−0.23	$\theta = 0.6$
			327	5×10^{-7} to 5×10^{-8}	−0.15	$\theta = 0.7$
	(111)	122	27	5×10^{-7} to 1×10^{-8}	−0.22	$\theta = 0.4$
Ir	(110)	123	27	$<10^{-6}$	−0.44	10 L
La	Film	74	27	$\sim10^{-5}$	+0.6	Saturated
	Film	74	−78	$\sim10^{-5}$	−1.0	Saturated, ΔSP becomes more positive with increasing P_{O_2}
	Film	74	−196	10^{-5}	−1.7	Saturated
Mg	Bulk	124		10^{-4}	+1.4	
	Abraded 3% Al alloy	39	25	7×10^{-9}	+1.5	6 min
Mn	Film	62	17	2×10^{-9}	0.0	2×10^{14} molecules cm^{-2}; "N" curve
Mo	Film	120	27	Doses	−1.8	$\theta = 1.3$
	Film	114	−195, 25	Doses 10^{-2}	−1.7	Saturated
	(110)	125	25	10^{-9} to 10^{-7}	−1.1	6 L
	(001)	125	25	10^{-9} to 10^{-7}	−1.2	6 L
	(100)	126	25	10^{-9} to 10^{-8}	−1.4	12 L, "N" curve
Ni	Film	114	−195	10^{-2}	−1.5	Saturated

TABLE 6.1. (*Continued*)

Metal	Form	Reference	T (°C)	P_{O_2} (Torr)	ΔSP (eV)	Remarks
Ni	Film	127	23	11	−1.25 to −0.75	Function of film sintering temperature (55–170°C)
	Film	114	25	10^{-2}	−0.95	Saturated
	(100)	66	29	10^{-8} to 10^{-6}	−0.3	1 L; $\Delta SP = +0.65$ at saturation
	(100)	66	−126	10^{-8} to 10^{-6}	−0.2	Saturated
	(001)	65	25	$\sim 2 \times 10^{-8}$ $c(2 \times 2)$	−0.36	
				$p(2 \times 2)$	−0.22	
	(110)	65	25	$\sim 2 \times 10^{-8}$ $(3 \times 1)(\frac{1}{3})$	−0.26	
				$(2 \times 1)(\frac{1}{2})$	−0.46	
				$(3 \times 1)(\frac{2}{3})$	−0.42	
	(111)	65	25	$\sim 2 \times 10^{-8}$ $p(2 \times 2)$	∼−0.7	
	(100)	128	25	$p(2 \times 2)$	−0.30	
				$c(2 \times 2)$	−0.37	
	(110)	128	25	(2×1)	−0.33	
				(3×1)	−0.44	
	(111)	128	25	(2×2)	−0.70	
Nb	(100)	61	25	2×10^{-8}	−1.6	6 L
	(110)	61	25	1×10^{-8} and 2×10^{-7}	−0.5 and −0.8	15 and 30 L, respectively
	(111)	61	25	6×10^{-8}	−1.2	8 L
Pb	Film	129	−195	6×10^{-3}	+0.31	Saturated
	Film	129	23	6×10^{-3}	+0.55	Saturated
Pt	Polysheet	130	22	5×10^{-9} to 5×10^{-7}	−0.44	5 L
	6 (111) × (111)	131	22	$\sim 10^{-8}$	−0.28	60 L
	6 (111) × (100)	131	22	$\sim 10^{-8}$	−0.47	50 L
Re	(0001)	132	25	3×10^{-8}	−0.9	8×10^{14} molecules cm^{-2}
	preferred orientation		577	3×10^{-8}	−1.3	1×10^{15} molecules cm^{-2}

Ru	(001)	133	−173	10^{-8} to 10^{-7}	−1.04	6 L
	(001)	133	27	10^{-8} to 10^{-7}	−0.82	6 L
Si	(100)	134	25	10^{-5}	+0.01	5 L; ΔSP = +0.27 at 1000 L
	(111)	134	25	10^{-5}	−0.67	5 L; ΔSP = −0.17 at 1000 L
Th	Film	135	25	5	+0.48	Saturated
	(100)	136	27	10^{-8}	+0.60	7 L; ΔSP = +0.45 at 36 L
Ti	Film	63	25	10^{-3}	−1.10	Saturated
	Film	120	27	Doses	−1.4	$\theta = 7.2$
	$(31\bar{4}0)$	137	100	2×10^{-8}	−1.23	23 min (28 L)
	Bulk	138	118	3×10^{-6}	−1.60	1000 sec
	$(31\bar{4}0)$	137	500	2×10^{-8}	−0.33	23 min (28 L)
U	Film	139	25	10	+0.68	Saturated
W	(100)	140	−195	Doses	−1.5	Saturated
		141	27	5×10^{-9}	−1.3	20 min
		142	25	10^{-7}	−1.5	10^{16} O_2 cm^{-2} exposure
		143	~27	O_2 beam	−1.6	10^{16} O_2 cm^{-2} exposure
		141	777	5×10^{-9}	−0.8	20 min.
		144	1560	4×10^{-7}	−0.94	$\theta = 1$; ΔSP = −1.7 at saturation
	(110)	145	25	WO_2 doses	−0.8	Saturated
		140	−195	Doses	−1.0	Saturated
		146	25	10^{-8} to 10^{-6}	−0.7	5 L
	(111)	140	−195	Doses	−1.8	Saturated
		147	−148	$\sim10^{-7}$	−1.7	10 L
		147	27	$\sim10^{-7}$	−1.3	10 L
	(112)	146	25	10^{-8} to 10^{-6}	−1.0	5 L
		148	25	$\sim2 \times 10^{-8}$	−0.9	$\theta = 0.8$
	(211)	140	−195	Doses	−1.3	Saturated

TABLE 6.2. SIZE OF ADSORPTION SITES ON THE (100) FACE [71]

Element	Pb	U	Al	Cu	Ni	Cr	Fe	Mo	W
Diameter of site (Å)	1.47	1.36	1.19	1.05	0.88	1.55	1.56	1.72	1.71
Depth of site (Å)	0.73	—	0.59	0.52	0.44	0.18	0.19	0.21	0.21

Oxygen adsorption up to a monolayer on a clean metal surface sometimes results in a decrease rather than an overall increase in work function. This has been explained in terms of oxygen fitting into interstices between the metal atoms. This model was advanced by Roberts and Wells [71] to account for the fact that oxygen adsorption on metals having large atomic radii, for example, Th or U, resulted in a decrease in work function. The spaces between atoms of such metals are large enough to admit oxygen ions having a radius of 1.40 Å as summarized in Table 6.2. This table helps show that the work function changes in Table 6.1 correlate with the size of the interstices in the metal crystal structure rather than with the size of the metal radii alone. The model is plausible since surface potential measurements detect the charged layer which is outermost on the metal–oxide surface. This may also account for the anomalous results obtained when water is present during low-temperature oxidation. The water dipoles may mask the field due to metal–oxygen charge separation in the oxide [72].

Heating a metal which is covered by an adsorbed layer of oxygen results in a decrease in work function because the oxygen is incorporated into the metal [72,73]. If oxygen is subsequently readsorbed on the metal, the work function again increases, but it does not necessarily return to the previous value.

Continued exposure of a stable oxide film to oxygen can result in an increase or decrease in work function, depending upon whether more oxygen ions can be accommodated on the oxide, creating an anion-rich surface, or whether cation incorporation into the oxide creates a cation-rich surface [74]. This process is temperature dependent as shown in Table 6.1. Fromhold [75] discusses the coupled electronic and ionic currents which control the value of surface potential. Basically, this potential adjusts to aid movement of the slower charged species as it traverses the oxide layer.

The exact point of transition from negative to positive surface potential change is not at all clear, as can be seen, for instance, from the scatter in aluminum data included in Table 6.1. EXAFS [76] has been used to show that oxygen penetrates below the top layer of aluminum at room temperature. The final value of surface potential depends on oxygen pressure, impurities, and condition of the metal surface. As pressure increases, the transition temperature for sign reversal can increase as in the case of lanthanum. Until all variables are controlled, surface potential measurements will continue to be more qualitative than quantitative [77].

6.3. CRYSTAL ORIENTATION EFFECTS

Thermal oxidation often depends on the crystallographic orientation of the metal, although in the cases of Ta or Al anodization, this dependence has not been found to hold true. Dignam [78] discusses some possible reasons for these different behaviors.

Different mechanisms are advanced to explain the variations in reactivity of individual crystallographic orientations of a metal, depending on the temperature range of oxidation. At low temperatures, the ionic and electronic currents are taken to be independent of each other so that the differing heats of solution of an ion in the oxide can account for the anisotropy in oxide growth rate [79]. This mechanism applies when thin uniform oxide films form.

A second explanation has been based on the differing atomic packing densities on the various crystallographic planes of the metal. This mechanism is thought to be particularly relevant when cations are mobile in the oxide, as is expected for modifying oxides. When anions are the mobile species, as proposed for network-forming oxides [72], then crystallographic dependence decreases as discussed below.

At intermediate temperatures, nucleation and growth of oxide crystallites can account for the differing reaction rates. The discussion on oxide islands in Chapter 5 indicates that initial formation of three-dimensional oxides relies on nucleation sites and surface diffusion of the reactants. Thus, the density and size of crystallites will be a function of substrate crystallography and perfection, as will the average oxide thickness.

Another effect of crystallographic orientation is found in the bulk oxide. Ion movement is controlled by the nature of the oxide. It turns out that the density of defects, such as low-angle grain boundaries which provide paths of easy-ion movement through the oxide, is again a function of substrate crystallography [80]. Although the differences in oxidation rate are not as striking at high temperatures as at low, the rate is found to vary with crystallographic orientation of the metallic substrate. Examples are given below.

Numerous studies have shown that epitaxy of oxide films is also related to the crystallographic orientation of the metal. Gwathmey and Lawless [81,82] made a detailed study of oxide orientation on copper for oxides formed at 150–350°C on various crystallographic faces of the metal. They

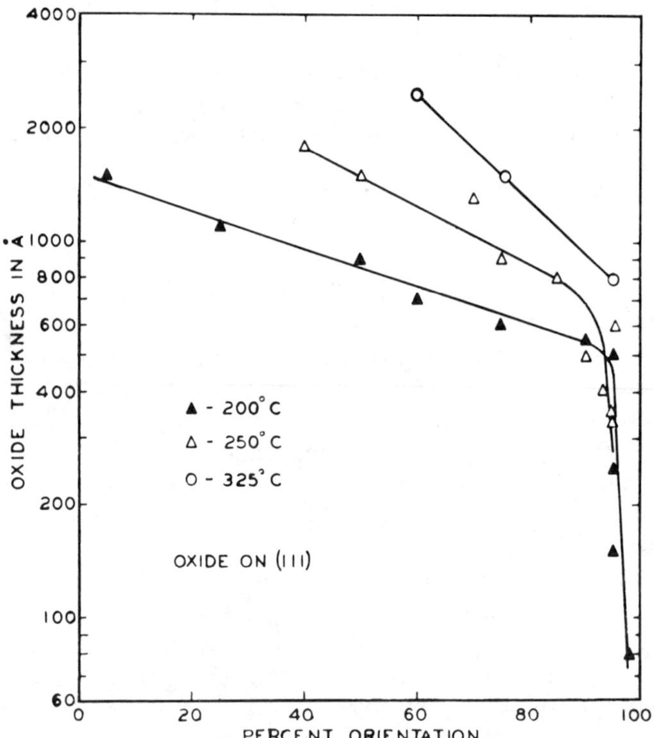

Figure 6.2. Variation of the degree of oxide orientation with thickness for Cu_2O on the (111) face of copper. Reprinted with permission from *Acta Metallurgica*, Vol. 4, p. 157, K. R. Lawless and A. T. Gwathmey, The Structure of Oxide Films on Different Faces of a Single Crystal of Copper, Copyright 1956, Pergamon Press, Ltd.

found that the oxide exhibited one or more orientations dependent on the copper face. Generally, it was observed that at least one close-packed direction in the oxide was parallel to a close-packed direction in the metal. A comparison with results of oxidation rate studies (Figure 6.2) also showed that the most reactive regions could have oxide with three or four orientations, and that the oxidation rates decreased as the number of oxide orientations decreased. This finding suggests a close correlation between nucleation, nuclei growth, and low-temperature oxidation.

The degree of oxide orientation was also a function of the reaction conditions and the oxide thickness. Thus, for copper oxidized at 350°C, oxide orientation decreased rapidly at oxide thicknesses above about 500 Å. The degree of orientation also became less marked with increasing temperature. Oxidation at reduced oxygen pressures favors the growth of oriented oxide. Other factors which affect oxide orientation include impurities and surface preparation [83]. Examples for several metals and one semiconductor are given below.

6.3.1. Copper

Copper has been extensively investigated. Rhodin [84] measured oxide thicknesses in the temperature range −195–50°C as a function of time using a microbalance as shown in Figure 6.3. Variables included oxygen pressure, temperature, and crystal orientation. It is immediately evident from Figure 6.3 that oxide thickness was a strong function of both temperature and crystal orientation. No pressure dependence was found by Rhodin. Cohen [85] found a similar lack of pressure dependence for magnesium.

The initial rate of copper oxidation decreased for the various crystal faces as (100) > (110) > (111). This order does not follow the packing density of nearest-neighbor copper atoms on various planes which is (110) < (100) < (111). Instead, it follows the order proposed by Tick and Witt [86] who considered this matter in terms of interactions between next-nearest as well as nearest neighbors. They would predict (100) > (110) > (111) for ease of atom removal from a particular crystal plane.

It is apparent from Figure 6.3 that the slope of oxide thickness versus the logarithm of time for long times is affected very little by crystal face and temperature, except at the highest temperature where a transition to cubic kinetics takes place.

Armitage et al. [87] also measured oxygen uptake on single crystal surfaces of copper at 50–60°C. Initial rates of uptake were (110) > (100) > (111). The difference was attributed to local atomic roughness on each crystal face, (110) being the roughest. The (311) face was expected to be slightly less reactive than the (110). Little difference in surface reactivity with copper orientation was noted after the initial chemisorption stage.

In the temperature range 70–178°C, Young, Cathcart, and Gwathmey [88] found that the reactivity decreased in the order (100) > (111) > (110) > (311). They took special care to ensure cleanliness after finding that oxidation of the (110) face was particularly sensitive to impurities. The increase in oxide thickness on the different faces as a function of time was measured by means of elliptically polarized light, and results at several temperatures are shown in Figure 6.4. After 50 min the oxide thickness on the (100) face was 12 to 13 times larger than that on the (311) face at 178°C.

6.3.2. Iron

The effect of orientation on the oxidation of iron single crystals has also been the subject of many studies. Wagner et al. [89] report rates of growth

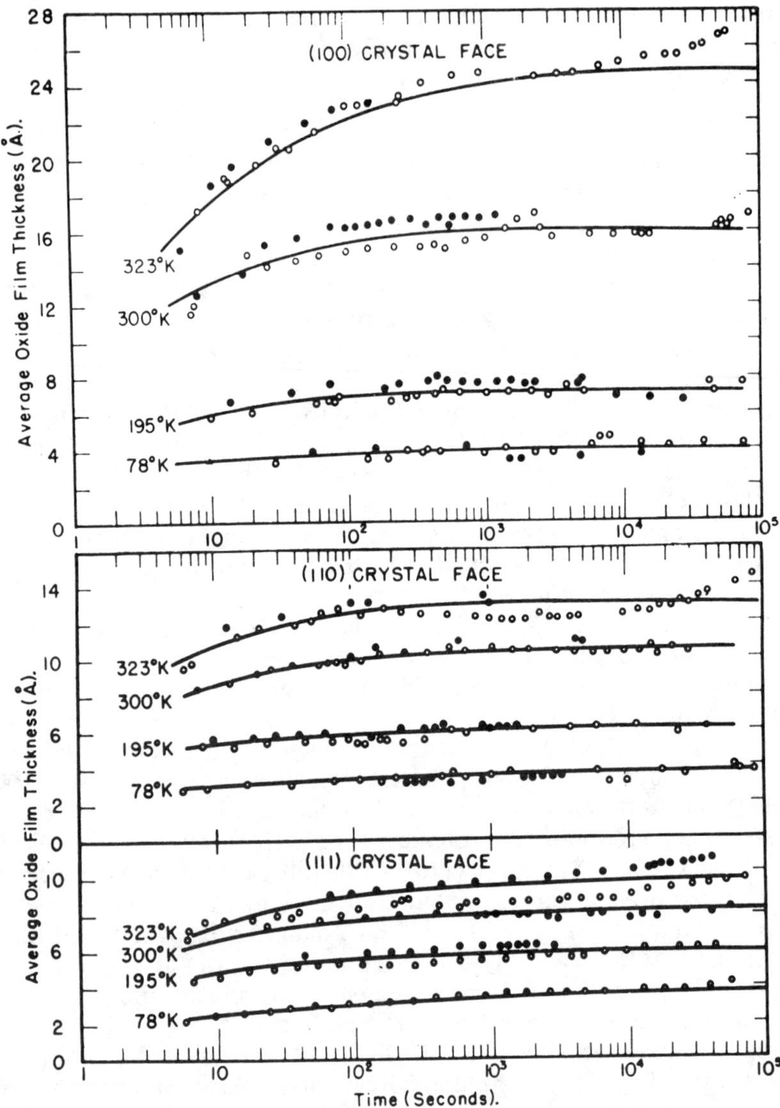

Figure 6.3. Oxidation of single-crystal copper surfaces as a function of temperature. Isotherm curves are calculated. Reprinted with permission from *J. Amer. Chem. Soc.*, Vol. 73, p. 3144, T. N. Rhodin, Low-Temperature Oxidation of Copper, Copyright 1951, American Chemical Society.

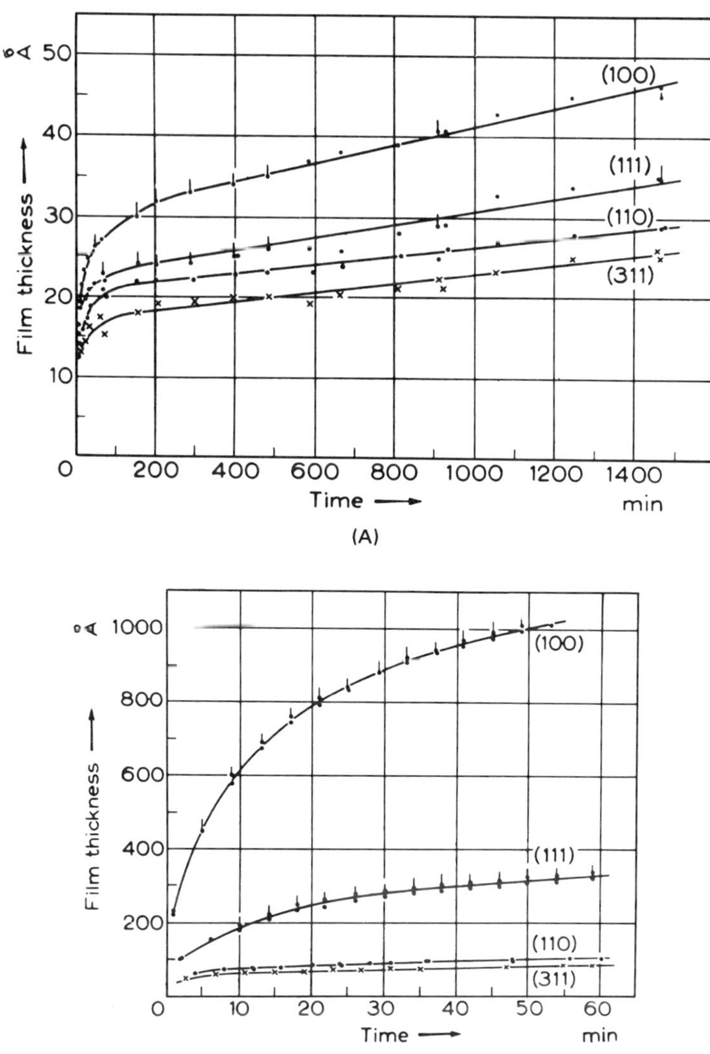

Figure 6.4. Oxidation of single crystal copper surfaces at (A) 70°C and (B) 178°C. Reprinted with permission from *Acta Metallurgica*, Vol. 4, p. 147, F. W. Young, Jr., J. V. Cathcart, and A. T. Gwathmey, The Rates of Oxidation of Several Faces of a Single Crystal of Copper as Determined with Elliptically Polarized Light, Copyright 1956, Pergamon Press, Ltd.

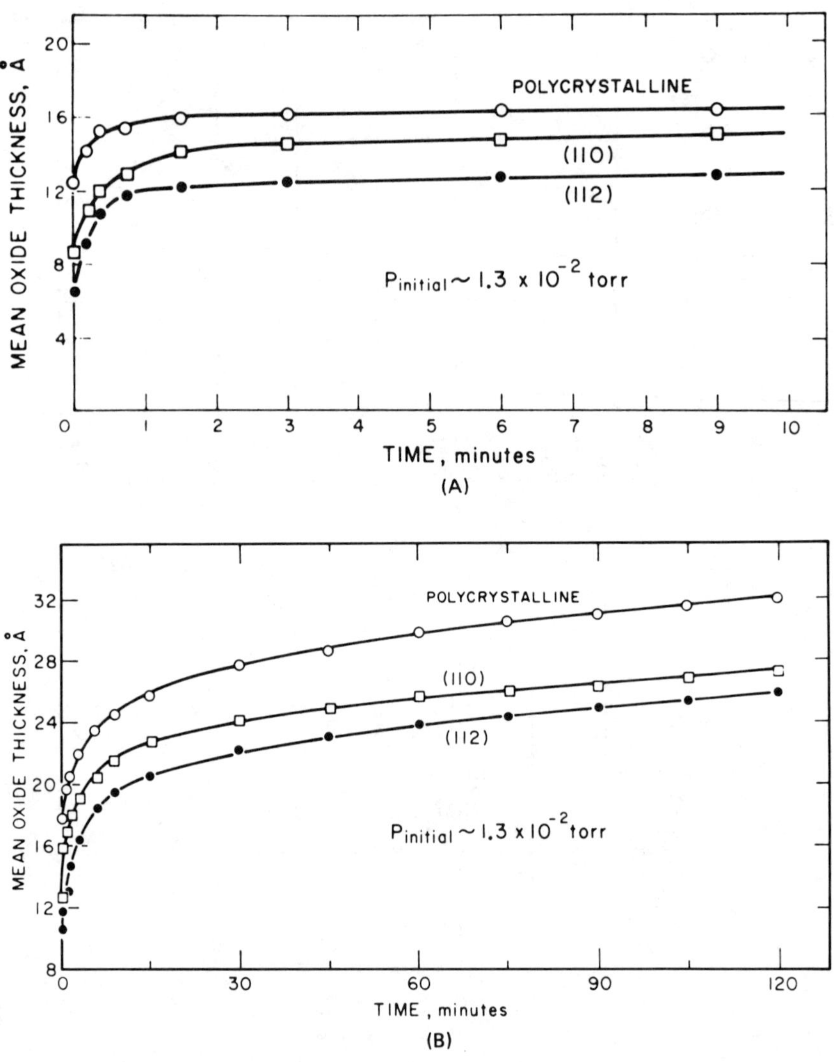

Figure 6.5. Oxidation of polycrystalline, (110), and (112) iron surfaces at (A) 24°C, (B) 100°C, and (C) 200°C. After Graham et al. [92]. Reprinted by permission of the publisher, The Electrochemical Society, Inc.

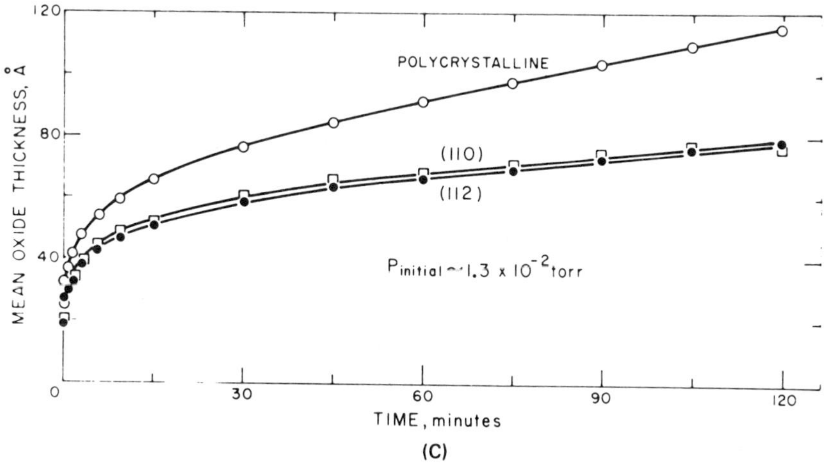

Figure 6.5. (*continued*)

between 250 and 550°C for 10–760-Torr pressure to be (110) > (111) > (110) > (320). The relative rates were found to change as a function of time. Predominant oxide orientation was

$$(001)_{Fe_3O_4} \parallel (001)_{\alpha-Fe} \qquad \text{and} \qquad [110]_{Fe_3O_4} \parallel [010]_{\alpha-Fe}$$

Sewell and Cohen [90,91] studied oxidation of iron single crystals around 200°C by means of electron diffraction, electrochemical, and gravimetric methods. They found that both the oxide composition and thickness were highly dependent on orientation [92]. Figure 6.5 shows the relative rates of oxidation at 24°C to be polycrystalline > (110) > (112). At 200°C the rates for the (110) and (112) faces of iron were almost identical.

Electron diffraction showed that at 200°C and below, the oxide consisted of only Fe_3O_4 growing epitaxially on the iron [91]. At oxidation temperatures of 260–400°C, α-Fe_2O_3 was also present. The evolution of the oxide composition is reviewed by Cohen [93]. The composition and thickness of the oxide film were also found to be dependent on the surface preparation and presence of impurities in the iron.

6.3.3. Nickel

Cohen's group has included nickel in their studies [94]. At 600°C the order of reaction was (100) > polycrystalline > (111) > (112) as shown in Figure 6.6. This result has been interpreted in terms of the number of easy dif-

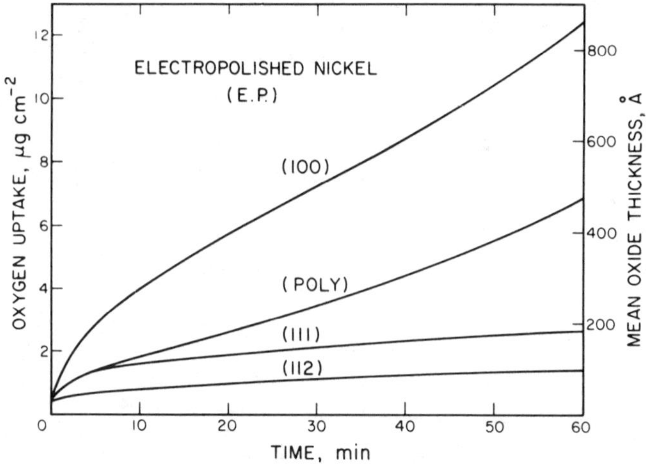

Figure 6.6. Oxidation curves at 600°C and 5×10^{-3} Torr oxygen for nickel crystals electropolished in sulfuric acid. After Graham et al. [94]. Reprinted by permission of the publisher, The Electrochemical Society, Inc.

fusion paths in the oxide, such as twin orientations or grain boundaries. The greater the number of defects, the higher the rate. Surface pretreatment affected the oxide structure and hence the rate. Hydrogen-fired surfaces free of S, Si, Fe, and C impurities oxidized faster than electropolished or etched nickel [95].

The results above appear to agree with the earlier work of Cathcart et al. [96]. They found that the oxidation rate at 500°C was qualitatively (110) > (100) > (111) > (311). Interpretation was based on short circuit diffusion paths.

Woodruff et al. [87,97] studied the initial oxidation of nickel at up to 300°C and found the relative rates to be (111) > (110) > (100). They discussed this result in terms of structural blocking of adsorption sites and the role of surface steps.

6.3.4. Silicon

Variation of oxidation rate with crystallographic orientation is small for silicon in dry oxygen [98,99] where (111) Si oxidizes slightly faster than (100) Si (800–1000°C, 1 atm). A reversal in these relative rates was found [100] at oxygen pressures below 0.1 atm. It was explained in terms of different crystallographic arrangements of silicon atoms at the Si/SiO_2 interface.

An interesting study of silicon oxidation in steam (500–800°C, 40–150 atm) has been carried out by Ligenza [101]. He separates the rates of oxidation, (110) > (310) > (111), into two terms, a preexponential factor and an activation energy. The preexponential agrees with the number of silicon atoms in the crystal surface while the activation energies are in the order (110) < (311) < (111). Silicon oxidation is discussed in more detail in Chapters 8 and 9.

6.4. OXYGEN PRESSURE DEPENDENCE

Oxygen pressure along with electronic flux controls the ion population at the oxide–gas interface and thereby the surface potential (Table 6.1) and the number of potentially mobile anions. Hence, those oxides in which anion movement predominates show a greater influence of oxygen pressure on the rate of low-temperature oxidation.

Examples such as aluminum [52,102] and silicon [103] produce oxides which are intermediate or network-forming oxides and in which anion movement is found. Hence, both N and V in the Cabrera–Mott expression for the rate of oxidation should be a function of pressure. This contrasts with modifying oxides where crystallographic orientation affects N, but not V. For anion movement, pressure may affect V more than N if part of the charge setting up the field across the oxide is fixed and thus unable to move through the oxide.

The functional dependence of the oxidation rate on pressure at low temperatures is still open to question. Fehlner [104] estimates a logarithmic pressure dependence for refractory metals in low-pressure dry oxygen while Roberts [105] reports a variety of dependencies including the square root of pressure. It is possible that impurities play a large role in affecting the surface dissociation and incorporation of oxygen and thus the pressure dependence of oxidation kinetics.

In summary, interfaces are a critical site of activity during oxidation. Cation incorporation or anion reaction occurs at the metal–oxide interface while anion incorporation or cation reaction occurs at the oxide–gas interface. Network-modifying oxides are expected to favor cation movement. As a result, oxidation kinetics should be crystallographically dependent provided ion incorporation is rate limiting. Network formers, on the other hand, are expected to show anion movement and oxygen pressure dependence. Experimental evidence supports this generalized view, but with some minor modifications yet to be defined

REFERENCES

1. B. Chatterjee, *Thin Solid Films* **41** (1977), 227.
2. M. Pepper, *Contemp. Phys.* **18** (1977), 423.

3. M. A. Heine and M. J. Pryor, *J. Electrochem. Soc.* **110** (1963), 1205.
4. M. J. Pryor, *Oxid. Metals* **3** (1971), 271, 523.
5. M. A. Heine, D. S. Keir, and M. J. Pryor, *J. Electrochem. Soc.* **112** (1965), 24.
6. M. A. Heine and P. R. Sperry, *ibid.* **112** (1965), 359.
7. G. Amsel, C. Cherki, G. Feuillade, and J. P. Nadai, *J. Phys. Chem. Solids* **30** (1969), 2117.
8. C. J. Dell'Oca, D. L. Pulfrey, and L. Young, in *Physics of Thin Films*, vol. 6, M. H. Francombe and R. W. Hoffman, Eds., Academic Press, New York, 1971, p. 1.
9. K. Shimizu, G. E. Thompson, and G. C. Wood, *Thin Solid Films* **85** (1981), 53.
10. L. Young, *Anodic Oxide Films*, Academic Press, New York, 1961, p. 68.
11. W. L. Lee, G. Olive, D. L. Pulfrey, and L. Young, *J. Electrochem. Soc.* **117** (1970), 1172.
12. J. D. Leslie and K. Knorr, *ibid.* **121** (1974), 263.
13. D. J. Smith and L. Young, *Thin Solid Films* **101** (1983), 11.
14. T. Ohtsuka and N. Sato, *J. Electrochem Soc.* **128** (1981), 2522.
15. A. T. Fromhold, Jr., *Thin Solid Films* **86** (1981), 57.
16. A. T. Fromhold, Jr., and N. Sato, *Phys. Lett.* **84A** (1981), 219.
17. T. W. Sigmon, W. K. Chu, E. Lugujjo, and J. W. Mayer, *Appl. Phys. Lett.* **24** (1974), 105.
18. R. Flitsch and S. I. Raider, *J. Vac. Sci. Technol.* **12** (1975), 305.
19. E. Taft and L. Cordes, *J. Electrochem. Soc.* **126** (1979), 131.
20. E. H. Nicollian and J. R. Brews, *MOS (Metal Oxide Semiconductor) Physics and Technology*, John Wiley & Sons, New York, 1982, p. 820.
21. H. R. Philipp, *J. Appl. Phys.* **43** (1972), 2835.
22. R. A. Clarke, R. L. Tapping, M. A. Hopper, and L. Young, *J. Electrochem. Soc.* **122** (1975), 1347.
23. R. Haight, W. M. Gibson, T. Narusawa, and L. C. Feldman, *J. Vac. Sci. Technol.* **18** (1981), 973.
24. A. C. Adams, T. E. Smith, and C. C. Chang, *J. Electrochem. Soc.* **127** (1980), 1787.
25. J. Derrien and M. Commandré, *Surface Sci.* **118** (1982), 32.
26. A. van Oostrom, L. Augustus, F. H. P. M. Habraken, and A. E. T. Kuiper, *J. Vac. Sci. Technol.* **20** (1982), 953.
27. C. R. Helms, in *Insulating Films on Semiconductors*, M. Schulz and G. Pensl, Eds., Springer-Verlag, Berlin, 1981, p. 19.
28. R. S. Bauer and R. Z. Bachrach, *J. Vac. Sci. Technol.* **17** (1980), 509.
29. N. M. Johnson, D. K. Biegelsen, and M. D. Moyer, *ibid.* **19** (1981), 390.
30. A. Many, Y. Goldstein, and N. B. Grover, *Semiconductor Surfaces*, North-Holland, Amsterdam, 1971.
31. M. M. Atalla, E. Tannenbaum, and E. J. Scheibner, *Bell Sys. Tech. J.* **38** (1959), 749.
32. L. Grunberg and K. H. R. Wright, *Proc. Roy. Soc. (London)* **A232** (1955), 403.
33. J. A. Ramsey and G. F. J. Garlick, *Brit. J. Appl. Phys.* **15** (1964), 1353.
34. F. R. Brotzen, *Phys. Stat. Sol.* **22** (1967), 9.
35. D. R. Arnott and J. A. Ramsey, *Surface Sci.* **28** (1971), 1.
36. C. Górecki, *J. Non-Cryst. Solids* **45** (1981), 63.
37. T. A. Delchar, *J. Appl. Phys.* **38** (1967), 2403.
38. N. Jakowski and H. Glaefeke, *Thin Solid Films* **36** (1976), 195.
39. T. F. Gesell and E. T. Arakawa, *Surface Sci.* **33** (1972), 419.
40. G. C. Allen, P. M. Tucker, B. E. Hayden, and D. F. Klemperer, *ibid.* **102** (1981), 207.
41. F. J. Morin and T. Wolfram, *Phys. Rev. Lett.* **30** (1973), 1214.
42. A. L. G. Rees, *Chemistry of the Defect Solid State*, Methuen, London, 1954.

43. H. E. Grenga and K. R. Lawless, *J. Appl. Phys.* **43** (1972), 1508.
44. E. G. Derouane, J. Fraissard, J. J. Fripiat, and W. E. E. Stone, *Catal. Rev.* **7** (1972), 121.
45. M. L. Hair, *Infrared Spectroscopy in Surface Chemistry*, Marcel Dekker, New York, 1967.
46. K. Siegbahn et al., *Electron Spectroscopy for Chemical Analysis*, Almqvist and Wiksells Boktryckeri AB, Uppsala, 1967.
47. Proc. Symp. Modern Methods of Surface Analysis, (Murray Hill, NJ, May 1970), *Surface Sci.* **25** (1971), 1–223.
48. D. O. Hayward and B. M. W. Trapnell, *Chemisorption*, Butterworths, London, 1964, p. 38ff.
49. R. V. Culver and F. C. Tompkins, *Advan. Catal. Relat. Subj.* **11** (1959), 67.
50. J. E. Inglesfield and B. W. Holland, in *The Chemical Physics of Solid Surfaces and Heterogeneous Catalysis*, D. A. King and D. P. Woodruff, Eds., Elsevier, New York, 1981, p. 183.
51. V. K. Agarwala and T. Fort, Jr., *Surface Sci.* **54** (1976), 60.
52. E. E. Huber, Jr., and C. T. Kirk Jr., *ibid.* **5** (1966), 447.
53. C. Benndorf, C. Nöbl, M. Rüsenberg, and F. Thieme, *Surface Sci.* **111** (1981), 87.
54. D. Ramanathan and P. Vijendran, in *Proc. 7th Intern. Vac. Congr. and 3rd Intern. Conf. Solid Surfaces*, R. Dobrozemsky et al., Eds., Vienna, 1977, p. 1101.
55. D. E. Grider, K. Bange, and J. K. Sass, *J. Electrochem. Soc.* **130** (1983), 246.
56. A. Spitzer and J. Lüth, *Surface Sci.* **120** (1982), 376.
57. F. P. Netzer and T. E. Madey, *ibid.* **127** (1983), L102.
58. C. T. Au, M. W. Roberts, and A. R. Zhu, *ibid.* **115** (1982), L117.
59. C-T. Au, J. Breza, and M. W. Roberts, *Chem. Phys. Lett.* **66** (1979), 340.
60. M. W. Roberts and P. R. Wood, *J. Electron. Spectrosc. Relat. Phenom.* **11** (1977), 431.
61. R. Pantel, M. Bujor and J. Bardolle, *Surface Sci.* **62** (1977), 589.
62. G. K. Hall and C. H. B. Mee, *ibid.* **28** (1971), 598.
63. W. Brearley and N. A. Surplice, *ibid.* **64** (1977), 372.
64. A. J. Melmed and J. J. Carroll, *J. Vac. Sci. Technol.* **10** (1973), 164.
65. J. E. DeMuth and T. N. Rhodin, *Surface Sci.* **45** (1974), 249.
66. P. H. Holloway and J. B. Hudson, *ibid.* **43** (1974), 123.
67. P. Hofmann, R. Unwin, W. Wyrobisch, and A. M. Bradshaw, *ibid.* **72** (1978), 635.
68. Y. Sakisaka, H. Kato, and M. Onchi, *ibid.* **120** (1982), 150.
69. H. Namba, J. Darville, and J. M. Gilles, *ibid.* **108** (1981), 446.
70. B. E. Hayden, E. Schweizer, R. Kötz, and A. M. Bradshaw, *ibid.* **111** (1981), 26.
71. M. W. Roberts and B. R. Wells, *Surface Sci.* **8** (1967), 453.
72. F. P. Fehlner and N. F. Mott, *Oxid. Metals* **2** (1970), 59.
73. M. W. Roberts and B. R. Wells, *Surface Sci.* **15** (1969), 325.
74. H. Heyne and F. C. Tompkins, in *Surface Phenomena of Metals*, S. C. I. Monograph No. 28, Society of Chemical Industry, London, 1968, p. 339.
75. A. T. Fromhold, *Theory of Metal Oxidation, Vol. I—Fundamentals,* North-Holland, New York, 1976.
76. M. L. den Boer, T. L. Einstein, W. T. Elam, R. L. Park, L. D. Roelofs, and G. E. Laramore, *J. Vac. Sci. Technol.* **17** (1980), 59.
77. J. Müller, *Surface Sci.* **69** (1977), 708.
78. M. J. Dignam, in *Oxides and Oxide Films*, Vol. 1, J. W. Diggle, Ed., Marcel Dekker, New York, 1972, p. 91.
79. N. Cabrera and N. F. Mott, *Rep. Prog. Phys.* **12** (1948–49), 163.

80. J. V. Cathcart, in *Oxidation of Metals and Alloys*, American Society for Metals, Metals Park, OH, 1971, p. 17.
81. A. T. Gwathmey and K. R. Lawless, in *The Surface Chemistry of Metals and Semiconductors*, H. C. Gatos et al., Eds., John Wiley & Sons, New York, 1960, p. 483.
82. K. R. Lawless and A. T. Gwathmey, *Acta Met.* **4** (1956), 153.
83. A. W. Swanson and H. H. Uhlig, *J. Electrochem. Soc.* **118** (1971), 1325.
84. T. N. Rhodin, Jr., *Advan. Catal.* **5** (1953), 39.
85. M. S. Cohen, *Acta Met.* **8** (1960), 356.
86. P. A. Tick and A. F. Witt, *Surface Sci.* **26** (1971), 165.
87. A. F. Armitage, H. T. Liu, and D. P. Woodruff, *Vacuum* **31** (1981), 519.
88. F. W. Young, Jr., J. V. Cathcart, and A. T. Gwathmey, *Acta Met.* **4** (1956), 145.
89. J. B. Wagner, Jr., K. R. Lawless, and A. T. Gwathmey, *Trans. Met. Soc. AIME* **221** (1961), 257.
90. P. B. Sewell and M. Cohen, *J. Electrochem. Soc.* **111** (1964), 501.
91. P. B. Sewell and M. Cohen, *ibid.*, 508.
92. M. J. Graham, S. I. Ali, and M. Cohen, *ibid.* **117** (1970), 513.
93. M. Cohen, *ibid.* **121** (1974), 191C.
94. M. J. Graham, R. J. Hussey, and M. Cohen, *ibid.* **120** (1973), 1523.
95. M. J. Graham, G. I. Sproule, D. Caplan, and M. Cohen, *ibid.* **119** (1972), 883.
96. J. V. Cathcart, G. F. Petersen, and C. J. Sparks, Jr., *ibid.* **116** (1969), 664.
97. H-T. Liu, A. F. Armitage, and D. P. Woodruff, *Surface Sci.* **114** (1982), 431.
98. E. A. Irene, *J. Electrochem. Soc.* **121** (1974), 1613.
99. L. N. Lie, R. R. Razouk, and B. E. Deal, *ibid.* **129** (1982), 2828.
100. S. I. Raider and L. E. Forget, *ibid.* **127** (1980), 1783.
101. J. R. Ligenza, *J. Phys. Chem.* **65** (1961), 2011.
102. C. T. Kirk, Jr. and E. E. Huber, Jr., *Surface Sci.* **9** (1968), 217.
103. Y. Kamigaki and Y., Itoh, *J. Appl. Phys.* **48** (1977), 2891.
104. F. P. Fehlner, *J. Electrochem. Soc.* **115** (1968), 726.
105. M. W. Roberts, *Quart. Rev. (London)* **16** (1962), 71.
106. H. A. Englehardt and D. Menzel, *Surface Sci.* **57** (1976), 591.
107. G. Rovida, F. Pratesi, M. Maglietta, and E. Ferroni, *ibid.* **43** (1974), 230.
108. V. K. Agarwala and T. Fort, Jr., *ibid.* **45** (1974), 470.
109. R. L. Wells and T. Fort, Jr., *ibid.* **33** (1972), 172.
110. P. Hofmann, W. Wyrobisch, and A. M. Bradshaw, *ibid.* **80** (1979), 344.
111. R. Michel, J. Gastaldi, C. Allasia, C. Jourdan, and J. Derrien, *ibid.* **95** (1980), 309.
112. V. K. Agarwala and T. Fort, Jr., *ibid.* **48** (1975), 527.
113. P. A. Anderson and A. L. Hunt, *Phys. Rev.* **115** (1959), 550.
114. C. M. Quinn and M. W. Roberts, *Trans. Faraday Soc.* **60** (1964), 899.
115. G. Gewinner, J. C. Peruchetti, A. Jaegle, and A. Kalt, *Surface Sci.* **78** (1978), 439.
116. C. Benndorf, B. Egert, G. Keller, H. Seidel, and F. Thieme, *J. Vac. Sci. Technol.* **15** (1978), 1806.
117. T. A. Delchar, *Surface Sci.* **27** (1971), 11.
118. A. Spitzer and H. Lüth, *ibid.* **118** (1982), 121.
119. A. Spitzer and H. Lüth, *ibid.*, 136.
120. J. Müller, *ibid.* **69** (1977), 708.
121. L. Surnev and M. Tikhov, *ibid.* **123** (1982), 505.
122. L. Surnev, *ibid.* **110** (1981), 439.
123. J. L. Taylor, D. E. Ibbotson, and W. H. Weinberg, *ibid.* **79** (1979), 349.
124. G. Blaise and G. Slodzian, *ibid.* **40** (1973), 708.
125. K. Hayek, H. E. Farnsworth, and R. L. Park, *ibid.* **10** (1968), 429.

126. R. Riwan, C. Guillot, and J. Paigne, *ibid.* **47** (1975), 183.
127. M. W. Roberts and B. R. Wells, *Trans. Faraday Soc.* **62** (1966), 1608.
128. G. E. Becker and H. D. Hagstrum, *Surface Sci.* **30** (1972), 505.
129. J. M. Saleh, B. R. Wells, and M. W. Roberts, *Trans. Faraday Soc.* **60** (1964), 1865.
130. D. M. Collins, J. B. Lee, and W. E. Spicer, *Surface Sci.* **55** (1976), 389.
131. D. M. Collins and W. E. Spicer, *ibid.* **69** (1977), 114.
132. J. Fusy, B. Bigeard, and A. Cassuto, *ibid.* **46** (1974), 177.
133. T. E. Madey, H. A. Engelhardt, and D. Menzel, *ibid.* **48** (1975), 304.
134. G. Hollinger and F. J. Himpsel, *J. Vac. Sci. Technol.* **A1** (1983), 640.
135. J. C. Rivière, *Brit. J. Appl. Phys.* **16** (1965), 1507.
136. T. N. Taylor, C. A. Colmenares, R. L. Smith, and G. A. Somorjai, *Surface Sci.* **54** (1976), 317.
137. J. B. Bignolas, M. Bujor, and J. Bardolle, *ibid.* **108** (1981), L453.
138. T. Smith, *ibid.* **38** (1973), 292.
139. J. C. Rivière, *Brit. J. Appl. Phys.* **15** (1964), 1341.
140. A. E. Bell, L. W. Swanson, and L. C. Crouser, *Surface Sci.* **10** (1968), 254.
141. E. Bauer, H. Poppa, and Y. Viswanath, *ibid.* **58** (1976), 517.
142. J. L. Desplat, *ibid.* **34** (1973), 588.
143. T. E. Madey, *ibid.* **33** (1972), 355.
144. A. E. Abey, *J. Appl. Phys.* **40** (1969), 284.
145. E. Bauer and T. Engel, *Surface Sci.* **71** (1978), 695.
146. J. C. Tracy and J. M. Blakely, *ibid.* **15** (1969), 257.
147. T. E. Madey, *ibid.* **29** (1972), 571.
148. D. L. Adams, L. H. Germer, and J. W. May, *ibid.* **22** (1970), 45.

7

FORMATION OF A STABLE OXIDE FILM

A characteristic feature of low-temperature oxidation for a large number of metals is an initially high rate of oxygen incorporation that quickly drops off to very low levels. It has been found that oxidation behavior can often be described in terms of the logarithmic-type equations discussed in Chapter 2.

A large number of theories and models have been put forward to explain the kinetics of low-temperature oxidation of metals, particularly the logarithmic oxidation behavior [1]. In many cases, it is difficult if not impossible to verify the correctness of the various models and the validity of parameters involved in the derived equations. Most of the theories assume the existence of uniform plane–parallel oxide films, but due to grain boundaries and defects, this assumption needs experimental verification.

The detailed chemical composition, that is, deviations from stoichiometry, of thin oxide films and the nature of defects in such films are difficult to study and hence poorly known at this time. Composition and defects may be a function of oxygen pressure, surface orientation, surface preparation, and impurities. Since thin film formation at low temperatures depends on rapidly changing systems under electrochemical driving forces where thermal diffusion is slow, equilibrium thermodynamic calculations of the defect concentrations in oxides are also questionable. For these reasons, low-temperature oxidation is still a relatively virgin field of investigation. Advances are being made, however, particularly through carefully controlled oxidation of metal single crystals.

It is assumed in many models that after an oxide film has covered a metal surface, either transport of electrons and ions through the film or reaction at an interface determines the reaction rate. At high temperatures, mass transport is assumed to involve diffusion driven by a chemical gradient across the oxide film, as described by the Wagner mechanism for parabolic oxidation. At low temperatures, an electric field across the film, resulting from chemisorption of oxygen ions on the oxide surface, is also taken into account to explain continued reaction. The role of stress in both high- and low-temperature oxidation is a real one, but currently poorly defined. It is expected to modify but seldom replace the chemical and electrical driving forces.

The multitude of oxidation models summarized by Kofstad [1] have validity under various restricted conditions. However, the Cabrera–Mott theory [2] has often been adopted to explain low-temperature oxidation [3]. The discussion below will emphasize the Cabrera–Mott treatment but will include modifications from other models [4,5,6,7,8,9,10] where appropriate.

7.1. CABRERA–MOTT MODEL

Immediately after the chemisorption stage of low-temperature oxidation, oxide islands increase in size until some of the ions become sluggish in moving by a place-exchange mechanism. As a result, anions remain for a longer and longer time on the oxide surface before place exchange and further oxide growth occur. Eventually, the ion population on the oxide surface increases to the point where ions no longer behave independently, so that lateral interactions between ions must be considered. The image force concept based on independent ions is then abandoned and the electric field based on the cooperative effect of many ions substituted. The field across the oxide, as given by Gauss' theorem for a field between parallel plates, becomes

$$E = V/x = \frac{\sigma''}{\epsilon} = \frac{Ne}{\epsilon_0 \kappa} \tag{1}$$

where σ'' is the areal charge density and N is the areal number density of anions. The potential V is generally 1 to 2 V. If, for example, $V = 0.3$, $x = 100$ Å, and $\kappa = 6$, then $N \cong 10^{12}$ cm^{-2}, much less than the number of cation sites ($\sim 10^{15}$ cm^{-2}) at the oxide–gas interface. Possible explanations for the smaller value of N involve limited impurity concentration on the oxide surface, special cation sites for oxygen adsorption on the oxide surface, kink or step sites at the metal–oxide interface, or channels which provide paths of easy-ion movement through the oxide.

The magnitude of the potential V is defined by the electronic levels in the metal and adsorbed oxygen. See Figure 2.2. Before electron transfer, the acceptor levels in the oxygen are below the Fermi level of the metal by the quantity eV. Electrons fill the oxygen levels until a quasisteady state of electrons is set up such that

$$eV = Y + W_b - \phi_v \tag{2}$$

where Y is the electron affinity of oxygen, W_b is the binding energy of an oxygen ion on the oxide surface, and ϕ_v is the work function of the metal against vacuum. Rough estimates of W_b allow one to estimate V. It is of the order of 1 V. Experimental surface potential values which are a partial measure of V help confirm this value. See Table 6.1.

In practice, V has been found to depend on oxygen pressure as shown in Table 6.1. This implies [6] that a distribution of energy states due to the adsorbed oxygen, rather than a single adsorption site, should be considered in calculating V. Such a distribution has been considered in the case of chemisorbed oxygen on a nickel cluster [11].

The field set up across an oxide of thickness x is taken to be V/x. This approximation ignores the possibility of space charge within the oxide and thus a concentration gradient of mobile species across the oxide, as pointed out by Hauffe [12a] and others [13]. Cabrera [14] has also cautioned that V never quite equals V_0, the equilibrium contact potential between metal and adsorbed oxygen, and that V/x decreases in a complicated way as x increases.

It is the field V/x, strong when x is small, that accounts for ion movement and hence continued oxide growth. This spontaneous, field-driven oxidation may occur momentarily during initial oxide growth on a clean metal surface at high temperatures, may continue to influence oxidation over a longer period of time at intermediate temperatures, and exists for measurable periods of time at low temperatures.

The mechanism of low-temperature oxidation is closely related to the mechanism of anodic oxidation, for in both cases high-field ionic movement is encountered in a homogeneous oxide. Many of the ideas expressed in the present work are taken from anodic studies and applied by analogy to low-temperature oxidation. The analogy is not exact due to the presence of water during anodization, with its consequent incorporation into the film structure [15,16]. However, the transport processes appear to be similar in low-temperature gas-phase oxidation, liquid-phase anodization, and gas-phase anodization [17,18]. Corrosion in liquid water to form either passive films or soluble corrosion products is similar to low-temperature oxidation in that an electrochemical potential is set up [19]. Insoluble products of corrosion differ from those of oxidation through incorporation of water to produce metal oxyhydroxides [20]. Solvation of reaction products must be taken into account when liquid water is present.

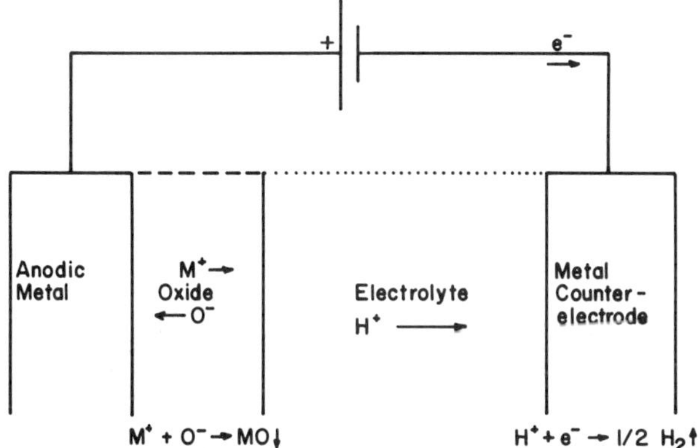

Figure 7.1. Schematic drawing of liquid-phase anodization. Equations at the interfaces only illustrate overall reactions.

The classical anodization of a metal is carried out in a liquid electrolyte as shown in Figure 7.1 so that a charged ion can be externally supplied to the surface of the growing oxide. Electronic conduction occurs in the external circuit. For this reason, only the movement of ions through the oxide under the influence of an applied field has been considered in formulating a mechanism for anodic oxide growth [21,22]. Even when electronic currents are present in the growing oxide, as in oxidation, they are not thought to interfere with ion movement.

However, during thermal oxidation, any movement of charged ions must be compensated by a simultaneous movement of another charged species, generally electronic. See Figure 7.2. In this way, overall charge neutrality is maintained. Since no external connections are made to the oxide as in anodization, it is obvious that countercurrents of charged species must move simultaneously through the oxide.

It was considerations of this kind that led Wagner [23], Hauffe [12b], Mott [24], Fromhold and Cook [25,26,27], Dignam [4,5], and others to consider the concept of coupled currents in which the sum of ionic and electronic currents must equal zero.

During the early stages of oxidation, as soon as an ion moves across the oxide to neutralize an ion of opposite charge (reacts to form oxide), the field across the oxide is momentarily reduced until an additional anion forms on the oxide surface. Anions form when electrons transfer from the metal to oxygen trap sites on the oxide surface (n-type oxide) or holes transfer from adsorbed oxygen to the metal (p-type oxide). In principle, electrons can transfer across an interface by thermionic emission or tun-

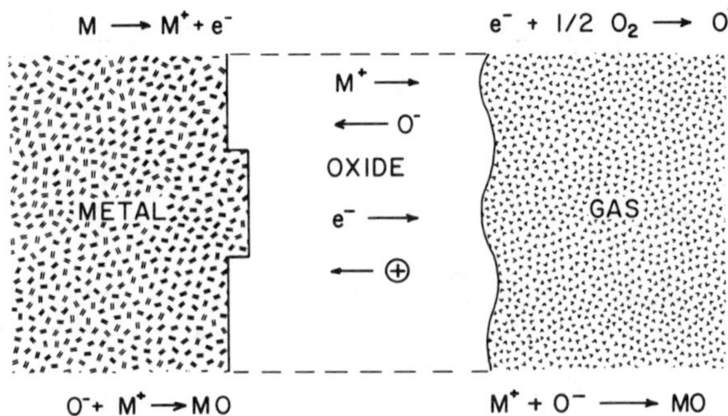

Figure 7.2. Schematic drawing of thermal oxidation. Equations at the interfaces only illustrate overall reactions.

neling. They can move through an oxide by band, impurity, or Poole–Frenkel conduction, or even tunnel completely through the oxide. The barrier to electron injection at a metal–oxide interface is of the order of 1 eV so that thermionic emission cannot occur easily at low temperatures. Tunneling, even when coupled with impurity conduction, is essentially independent of temperature and can account for the easy availability of electrons at the oxide–gas interface. See Figure 2.6.

The process which produces anions and the reaction to form oxide remain in virtual equilibrium during the transition from a clean metal surface to three-dimensional oxide. The concentration of anions is therefore constant for a particular set of conditions, for example, oxygen pressure and temperature. There is some evidence implying that anion concentration compensates to maintain a constant field across the oxide [28] during initial oxide growth. For instance, the pressure dependence of the surface potential of aluminum is shown in Figure 7.3. An increase in oxygen pressure raises the surface potential. To compensate for the increase in surface potential and thus maintain a constant field during the transition period, the oxide thickness also increases. After equilibrium is reestablished, the surface potential remains almost constant as film thickness continues to increase. The field then is postulated to decrease during the period of logarithmic oxide growth.

The anion surface concentration on aluminum exposed to oxygen is temperature dependent, as found in the surface potential work of Roberts and Wells [29]. Figure 7.4 shows that, as oxygen is added at −195°C, additional anions can form. Heating causes a decrease in anion population, but not through desorption of adsorbed oxygen. Instead, oxygen is

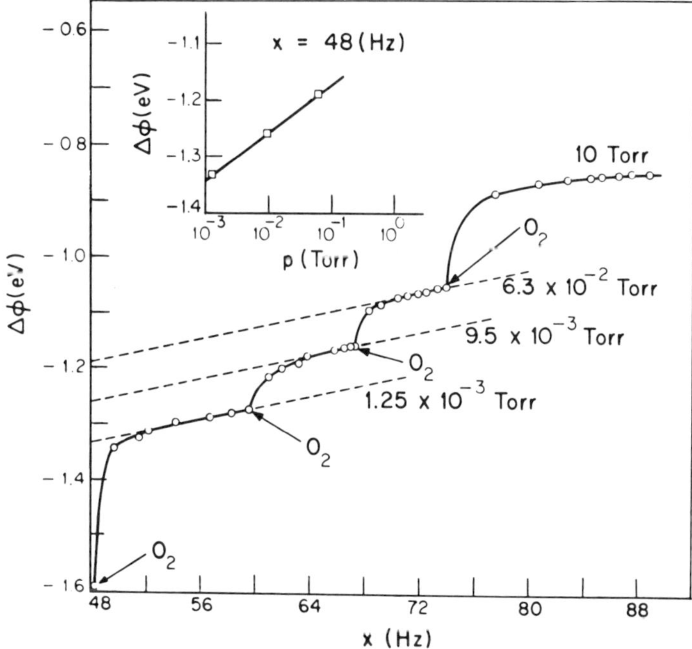

Figure 7.3. Work function difference as deduced from the contact potential difference between an oxidizing aluminum film and a gold reference electrode versus oxide thickness as measured with a quartz crystal microbalance having a sensitivity of 4.7×10^{-9} g cm^{-2} Hz^{-1} (0.28 Å of bulk Al_2O_3) as a function of pressure. After Kirk and Huber [28]. Published by permission of the copyright holder, North-Holland Physics Publishing.

incorporated into the metal through solution or anion–cation reaction to form additional oxide.

The limiting process giving rise to low-temperature oxidation could in principle be either an electronic or ionic one [30]. Fromhold [27] has analyzed these possible rate-controlling steps in a general treatment of the coupled-currents approach to oxidation. See Section 2.10. While it is possible, as in halide salts, to have electronic rate control and in fact, Mott [24,31] suggested it quite early for oxidation at low temperatures, there is still little experimental proof that it exists for such oxide growth. The electron model proposed by Mott assumed that tunneling through the oxide film is rate determining. On this basis, he derived a direct logarithmic rate equation for the growth of oxide films. It can be expressed as

$$x = k_1 \log t + k_2 \tag{3}$$

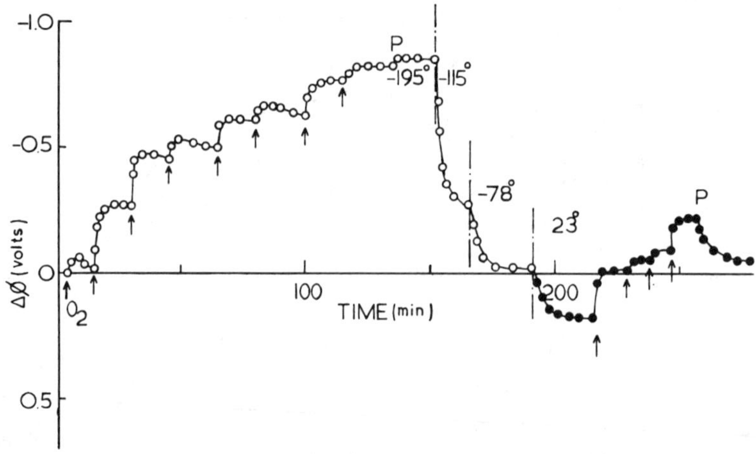

Figure 7.4. Change in work function of an aluminum film upon interaction with oxygen at −195°C followed by evacuation (*P*), step-wise heating to 23°C, and oxygen adsorption at 23°C. The arrows indicate the time of admission of each oxygen dose. After Roberts and Wells [29]. Published by permission of the copyright holder, North-Holland Physics Publishing.

Hauffe and Ilschner [32] have further elaborated on the model. Evans [33] has given an alternative derivation employing the same basic assumption.

The work of Miles [34] on solid-state anodization indicates that a virtual electronic equilibrium may be usual for thin oxides at low temperatures. Miles found that, even when an external connection was used to short circuit the electronic current, the rate of oxide growth was comparable to that found for thermal oxidation where internal electronic currents compensated for ion movement.

The electrochemical oxidation of nickel and iron at temperatures above 350°C was carried out by Yang and O'Grady [35] using a solid oxide electrolyte. They found close similarities between the thermodynamic properties determined in this way and those determined in gas-phase oxidation. Dignam et al. [36] conclude that a virtual electronic equilibrium is frequently attained during low-temperature oxidation. On the other hand, the low-temperature oxidation of tantalum has been interpreted in terms of electron tunneling [37].

Calculations by Fromhold and Cook [26] show that electron transfer in a one-step tunneling process occurs easily up to an oxide thickness of ~30 Å. Electron transfer by impurity conduction [38] at low temperatures is also possible for oxide thicknesses greater than 30 Å. We assume then,

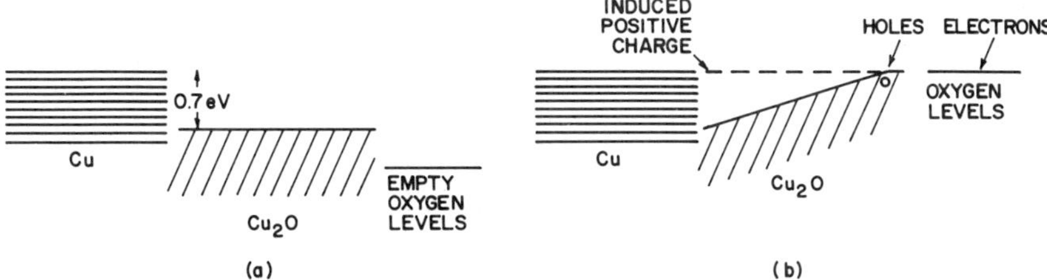

Figure 7.5. Electronic levels in Cu, Cu_2O, and oxygen: (a) before equilibrium is reached and (b) afterwards. After Cabrera and Mott [2]. Published by permission of the copyright holder, The Institute of Physics.

for our initial discussion, that the ionic processes which occur during low-temperature oxidation are rate limiting as proposed by Cabrera and Mott [2] and reiterated by Fehlner and Mott [6].

The model as stated in Chapter 2 is that electrons tunnel to the surface oxygen traps until sufficient traps are full to equalize the Fermi levels on both sides. The traps may empty by tunneling of electrons back to the metal, but the supply of electrons is sufficient to ensure that the traps are kept full in a quasisteady state. Electron tunneling is assumed to account for electron transfer in oxides where the trap level occurs in the forbidden band of the oxide. However, hole movement is also possible. Cabrera and Mott [2] considered the case of copper oxide where the trap level is taken to be below the valence band of the oxide. In such a situation, charging of the traps occurs by hole transfer via the valence band of the oxide as shown in Figure 7.5. As in the case of electron transfer, the field across the oxide is given by Equation 1. It is this field which lowers the activation energy for ion movement and thereby brings about oxide growth at low temperatures.

The electronic and ionic currents simultaneously present in an oxide may be considered separately if the effects of space charge can be ignored. Following the treatment of Debye and Hückel, Cabrera and Mott showed that ions and electrons neutralize each other for distances greater than the Debye length. For instance, values of the Debye length for oxides at room temperature are of the order of 0.1–1 μm, a number greater than the thickness of oxide formed at that temperature. Hence, for temperatures below 300–500°C, depending on the oxide, one may consider the electronic and ionic currents separately as in anodization. This is a great convenience as we shall see below.

A direct result of this separation of currents is the possibility of explaining the crystallographic dependence of low-temperature oxidation during the logarithmic growth period. The heat of solution of a metal in its oxide is $\phi + W_i - \gamma$ where W_i is the heat of solution of an ion in the oxide, γ is the energy with which an electron is bound to the interstitial ion in the oxide, and ϕ is the energy required to promote an electron from the metal into the conduction band of the oxide. When complete dissociation occurs, γ is small and may be ignored.

It should be noted that ϕ and W_i individually depend on which crystal face of the metal is being considered. However, the sum $\phi + W_i$ is not crystallographically dependent because it represents the heat of solution of any metal atom in the oxide. When ionic and electronic processes are considered separately, then the crystallographic dependence of the oxidation process may be explained by the differences in W_i. In the case of thicker oxides formed at higher temperatures, crystallographic dependence of oxidation can no longer be explained in this way. Instead, nucleation of islands at preferred sites [39] and paths of easy-ion movement [40] are considered to explain the differences. This is discussed in more detail in the section on interface effects. See Chapter 6.

In summary then, the Cabrera–Mott theory assumes that oxygen atoms are adsorbed on the oxide surface and that electronic species can pass rapidly through the oxide by tunneling to establish an equilibrium between the metal and adsorbed oxygen. This process creates anions and thus an electric field which facilitates the transport of ions across the oxide film. Dissolution of cations at the metal–oxide interface or anions at the oxide–gas interface is assumed to be rate determining, instead of ion movement in the oxide itself. Space charge in the oxide film is neglected. Under these conditions it can be shown (Equations 28, 29 and 30 of Chapter 2) that oxide film growth can be represented by the expression

$$\frac{x_1}{x} = -\ln\left(\frac{t}{x^2}\right) - \ln(x_1 u) \tag{4}$$

which is termed the inverse or reciprocal logarithmic rate law. It applies when either cations or anions are mobile and has an implicit pressure dependence in the value of N, the number of potentially mobile ions. Temperature dependence occurs in both N and W/kT. Detailed application of the Cabrera–Mott theory [4,27,41] takes these variables into account.

Hauffe [12,42] has criticized the Cabrera–Mott assumption that the defect concentration is constant within the oxide film. He has shown, however, that the same rate equation can be derived without making this assumption. Expressions similar to Equation 4 have also been derived by Evans [33] and Hurlen [43] from general oxidation equations assuming rate-limiting ion transport under the action of an electrical potential gradient.

7.2. ALTERNATIVE MODELS

Grimley and Trapnell [41] have considered possible mechanisms for thin film oxidation using the model of Mott and Cabrera, but assuming a constant field intensity in the oxide rather than a constant potential difference across the oxide. On this basis, linear rate equations were obtained if surface reactions or a transport of metal ions in interstitial positions were rate determining, and a direct logarithmic equation was obtained if the transport of metal ions took place via vacancies. Hauffe [12] discusses the evidence supporting the constant field assumption.

Logarithmic rate equations have been derived by Landsberg [44] and Halsey [45] on the assumption that chemisorption is rate determining. Landsberg bases his approach to oxidation on the blocking and generation of surface sites for oxygen adsorption while Halsey considers adsorption on a nonuniform catalyst surface.

In other treatments of low-temperature oxidation of metals, the effects of space charge in the oxide films are taken into consideration. Uhlig [46] has derived a dual logarithmic rate equation assuming that the oxidation rate is controlled by electron flow from the metal to the oxide. He suggests that the electron flow is a function of space charge set up near the metal–oxide interface. The space charge changes from uniform to nonuniform with time as increasing numbers of electrons become trapped at oxide defect sites. Fromhold's comment [47] on Uhlig's conclusion and Uhlig's reply [48] have been modified by Chattopadhyay [49] and applied to oxidation by Chatterjee [50]. Another modification of Uhlig's model has been proposed by Williams and Hayfield [51]. In both cases, a single space charge region is assumed and a direct logarithmic rate law is derived. Dignam [4] and Fromhold [27] have further considered the effects of space charge in oxide films. Engell, Hauffe, and Ilschner [52] have derived a cubic rate equation based on space charge effects. It is applicable to *p*-type oxide films.

Evans [33] and Davies, Evans, and Agar [53] have proposed models for which transport of reacting atoms or ions is assumed to take place preferentially along pores or low-resistance diffusion paths, that is, grain boundaries, dislocations, or channels. These models, based on self- or mutual blockage of the paths of easy-ion movement (see Section 2.3) lead to logarithmic type rate laws. Harrison [54] has also discussed simultaneous self-blockage and mutual blockage.

Anderson and Gallagher [55] considered the case where reactions are highly exothermic such that microscopically localized heating may cause enhanced solid-state diffusion. When oxygen is incorporated into the oxide lattice and the rate of heat dissipation is small, the oxygen atom may retain an appreciable fraction of the excess energy and behave as a hot atom which can diffuse appreciable distances in the solid. If the oxygen atom can reach different sites and if the different jump processes involve a wide distribution of activation energies, then the total reaction may follow the logarithmic rate equation. On the other hand, if the kinetic energy of the reacting oxygen atoms is rapidly transferred to the crystal lattice, a local temperature rise occurs in the surface layer. Depending on whether or not a thermal steady state is reached, the rate of reaction would decrease or increase with time.

It may be noted that the concept of independent steps having a distribution of activation energies as used by Anderson and Gallagher is basically the same as that in the Elovich equation used to describe chemisorption kinetics. Kofstad [56] used this model to interpret logarithmic oxidation of polycrystalline tantalum. It was assumed that nucleation and growth of tantalum suboxides varied over the surface and were dependent on surface orientation, defects, and grain boundaries. If a distribution of activation energies were involved, then a logarithmic rate equation would be obtained.

A process which gives rise to logarithmic kinetics is the lateral growth of three-dimensional oxide islands on a metal or thin oxide-covered metal surface. Evans' [33] treatment of this problem is summarized in Chapter 2 for both constant nuclei density and constant nucleation rate. If nuclei also grow simultaneously at a linear rate normal to the surface by a rate-controlling phase-boundary process, the reaction rate will initially increase and eventually become linear when the surface is completely covered. Alternatively, if the oxide growth normal to the surface is parabolic due to diffusion control, the oxidation rate will initially increase and eventually decrease according to a parabolic rate. The latter type of kinetics is described by a sigmoidal curve. Bartlett [57] treats these models for the case of aluminum.

The role of stress in low-temperature oxidation is a complex subject and one that has been little explored. The growth of oxide films on metals can lead to either compressive or tensile stresses, dependent upon oxide properties and the mode of growth. Pilling and Bedworth [58] early pointed out that the difference in molar volumes between the metal and oxide would certainly lead to stress. However, their ideas have been extensively modified by later authors. Stress relief occurs when dislocations are formed, provided that the mismatch between metal and oxide is greater than some minimum value, ~9% according to van der Merwe et al. [59,60] for epitaxial growth. Stress can also be relieved by the growth of a vitreous oxide which relaxes through viscous flow to achieve almost complete bonding at the boundary between the metal and oxide.

The mode of ionic transport is important in stress generation. Anion movement results in oxide formation at the metal–oxide interface where mechanical constraints can lead to stress. On the other hand, cation movement results in oxide growth at the free interface between oxide and oxidant where stress can be relieved.

Bradhurst and Leach [61], in qualitative agreement with Vermilyea's work [62], show that a high growth rate for anodic oxide on aluminum produces tensile stress, while slow growth results in compressive stress. It is known from the work of Davies et al. [63] that cation movement increases relative to anion movement as the growth rate (current density) in 3% aqueous ammonium citrate is increased. Thus, less oxide grows at the oxide–metal interface and the stress may shift from compressive to tensile. The role of water in this process is also important [62].

A more detailed presentation of stress effects is given by Fromhold [64]. He discusses the interaction of mobile anionic and cationic species with the supporting oxide. Energy exchange creates heat. The observation that the diffusing species move with constant velocity means that the diffusion medium exerts an opposing force on the mobile species. This force is equivalent to the applied force. Even when coupled currents are considered, a net force is experienced which creates stress in the oxide. This agrees with the experimental work of Borie, Sparks, and Cathcart [65] on the oxidation of copper.

Fromhold [64] also points out the possible role of diffusing defects in causing stress, provided the diffusion medium cannot relax to accommodate the defect. He calculates the effect of a voltage V across the oxide film. Thus, longitudinal compressive stress χ_i across an oxide of thickness L would be

$$2\chi_i = \frac{-\kappa V^2}{4\pi L^2} = -4\pi \frac{(q\sigma'')^2}{\kappa} \tag{5}$$

where σ'' is the surface charge density. For instance, 0.25 V across an oxide 250 Å thick generates a stress of 1.6×10^5 dyne/cm^{-2} or 2.32 lb in.$^{-2}$, a relatively large value.

The presence of such large stresses in oxide films on metals may help explain phenomena noted during the growth of vitreous oxides by anodic and low-temperature thermal oxidation. Bradhurst and Leach [66], Leach and Neufeld [67], and others have noted that during anodization, aluminum oxide films show a plasticity that is not present after growth ceases. The effect is ascribed to movement of ions from regions of compression to regions of tension. The ability to deform an oxide-coated aluminum wire without cracking the oxide was found to be proportional to both applied stress and an electric field applied across the oxide. It was concluded that plasticity of the oxide was increased by both stress and field. The temperature of the aluminum wire increased ~50°C during anodization at 200 mA cm^{-2}, but this effect was distinguished from stress in estimating the increased ductility of the oxide. Ord et al. [68,69,70] have included the role of stress in their interpretation of ellipsometric studies of anodic oxide growth. Further discussion of this topic is found in Chapter 2.

The model of Cabrera and Mott assumes that a uniform oxide is formed. In practice, one of three types of oxide may form, vitreous, microcrystalline, or polycrystalline. The first two appear to fit the Cabrera–Mott model as discussed below, but the third represents an exception. The existence of grain boundaries provides paths of easy oxidant movement, leading to a situation where transport down grain boundaries may become dominant. This subject is reviewed by Cathcart [40].

In summary, the evolution of an adsorbed layer of oxygen into a three-dimensional oxide film can be explained according to several mechanisms. The Cabrera–Mott model appears to be well founded and will be tested in Chapter 8.

REFERENCES

1. P. Kofstad, *High–Temperature Oxidation of Metals*, John Wiley & Sons, New York, 1966, p. 46.
2. N. Cabrera and N. F. Mott, *Rept. Prog. Phys.* **12** (1948–49), 163.
3. D. J. Young and M. Cohen, *J. Electrochem. Soc.* **124** (1977), 769.
4. M. J. Dignam, in *Oxides and Oxide Films*, Vol. 1, J. W. Diggle, Ed., Marcell Dekker, New York, 1972, p. 91.
5. M. J. Dignam, in *Comprehensive Treatise of Electrochemistry*, Vol. 4, J. O'M. Bockris, B. E. Conway, E. Yeager, and R. E. White, Eds., Plenum Press, New York, 1981, p. 247.
6. F. P. Fehlner and N. F. Mott, *Oxid. Metals* **2** (1970), 59.
7. M. J. Pryor, *ibid.* **3** (1971), 271.
8. F. P. Fehlner and N. F. Mott, *ibid.*, 275.
9. A. K. Vijh, *ibid.* **4** (1972), 63, 79.
10. F. P. Fehlner and N. F. Mott, *ibid.*, 75.
11. I. P. Batra and O. Robaux, *Surface Sci.* **49** (1975), 653.
12. K. Hauffe, *Oxidation of Metals*, Plenum Press, New York, 1965: (a) p. 125, (b) p. 129ff.
13. J. G. Simmons, in *Handbook of Thin Film Technology*, L. I. Maissel and R. Glang, Eds., McGraw-Hill, New York, 1970, p. 14-1ff.
14. N. Cabrera, in *Semiconductor Surface Physics* (Proc. Conf. Physics Semicond. Surf., Philadelphia, 1956), R. H. Kingston, Ed., University of Pennsylvania Press, Philadelphia, 1957, p. 327.

15. C. J. Dell'Oca, D. L. Pulfrey, and L. Young, in *Physics of Thin Films*, Vol. 6, M. H. Francombe and R. W. Hoffman, Eds., Academic Press, New York, 1971, p. 1.
16. R. W. Revie, B. G. Baker, and J. O'M. Bockris, *J. Electrochem. Soc.* **122** (1975), 1460.
17. W. L. Lee, G. Olive, D. L. Pulfrey, and L. Young, *ibid.* **117** (1970), 1172.
18. W. D. Mackintosh and H. H. Plattner, *ibid.* **124** (1977), 396.
19. H. H. Uhlig, *Corrosion and Corrosion Control*, 2nd ed., John Wiley & Sons, New York, 1971.
20. *Passivity of Metals* (Proc. 4th Intern. Symp. Passivity), R. P. Frankenthal and J. Kruger, Eds., The Electrochemical Society, Princeton, NJ, 1978.
21. M. Croset, E. Petreanu, D. Samuel, G. Amsel, and J. P. Nadai, *J. Electrochem. Soc.* **118** (1971), 717.
22. J. Siejka, J. P. Nadai, and G. Amsel, *ibid.* **118** (1971), 727.
23. C. Wagner, *Z. physik. Chem.* **B21** (1933), 25.
24. N. F. Mott, *Trans. Faraday Soc.* **36** (1940), 472; **43** (1947), 429.
25. A. T. Fromhold, Jr., *J. Phys. Chem. Solids* **33** (1972), 95.
26. A. T. Fromhold, Jr. and E. L. Cook, *J. Appl. Phys.* **38** (1967), 1546; *Phys. Rev.* **158** (1967), 600; **163** (1967), 650.
27. A. T. Fromhold, Jr., *Theory of Metal Oxidation, Vol. I—Fundamentals*, North-Holland, New York, 1976.
28. C. T. Kirk, Jr. and E. E.Huber, Jr., *Surface Sci.* **9** (1968), 217.
29. M. W. Roberts and B. R. Wells, *ibid.* **8** (1967), 453; **15** (1969), 325.
30. R. B. Mosley and A. T. Fromhold, Jr., *Oxid. Metals* **8** (1974), 19, 47.
31. N. F. Mott, *Trans. Faraday Soc.* **35** (1939), 1175.
32. K. Hauffe and B. Ilschner, *Z. Elektrochem.* **58** (1954), 467.
33. U. R. Evans, *The Corrosion and Oxidation of Metals*, St. Martins Press, New York, 1960.
34. J. L. Miles, *Proc. Roy. Soc. (London)* **A321** (1971), 503.
35. C. Y. Yang and W. E. O'Grady, *J. Vac. Sci. Tech.* **20** (1982), 925.
36. M. J. Dignam et al., *J. Phys. Chem. Solids* **34** (1973), 1227, 1235.
37. P. B. Sewell, D. F. Mitchell, and M. Cohen, *Surface Sci.* **29** (1972), 173.
38. N. F. Mott and E. A. Davis, *Electronic Processes in Non-crystalline Materials*, 2nd ed., Clarendon Press, Oxford, 1978.
39. J. Bénard, in *Oxidation of Metals and Alloys*, D. L. Douglass, Ed., American Society for Metals, Metals Park, OH, 1971, p. 1.
40. J. V. Cathcart, *ibid.*, p. 17.
41. T. B. Grimley and B. M. W. Trapnell, *Proc. Roy. Soc. (London)* **A234** (1956), 405.
42. K. Hauffe, in *The Surface Chemistry of Metals and Semiconductors*, H. C. Gatos, Ed., John Wiley & Sons, New York, 1960, p. 439.
43. T. Hurlen, *Acta Chem. Scand.* **13** (1959), 695.
44. P. T. Landsberg, *J. Chem. Phys.* **23** (1955), 1079.
45. G. D. Halsey, Jr. *J. Phys. Colloid Chem.* **55** (1951), 21.
46. H. H. Uhlig, *Acta Met.* **4** (1956), 541.
47. A. T. Fromhold, Jr., *J. Electrochem. Soc.* **115** (1968), 882; *Nature (London)* **200** (1963), 1309.
48. H. H. Uhlig, *J. Electrochem. Soc.* **116** (1969), 435.
49. B. Chattopadhyay, *Thin Solid Films* **16** (1973), 117.
50. B. Chatterjee, *ibid.* **35** (1976), 397; **41** (1977), 227.
51. E. C. Williams and P. C. S. Hayfield, in *Vacancies and Other Point Defects in Metals and Alloys*, Monograph No. 23, The Institute of Metals, London, 1958, p. 131.
52. H. J. Engell, K. Hauffe, and B. Ilschner, *Z. Elektrochem* **58** (1954), 478.

53. D. E. Davies, U. R. Evans, and J. N. Agar, *Proc. Roy. Soc.* (*London*) **A225** (1954), 443.
54. P. L. Harrison, *J. Electrochem. Soc.* **112** (1965), 235.
55. J. S. Anderson and K. J. Gallagher, in *Reactivity of Solids* (Proc. 4th Intern. Symp. Reac. Solids, Amsterdam, 1960), J. H. deBoer, Ed., Elsevier, New York, 1961, p. 222.
56. P. Kofstad, *J. Inst. Metals* **91** (1962–63), 209.
57. R. W. Bartlett, *J. Electrochem. Soc.* **111** (1964), 903.
58. N. B. Pilling and R. E. Bedworth, *J. Inst. Metals* **29** (1923), 529.
59. J. H. van der Merwe and N. G. Van der Berg, *Surface Sci.* **32** (1972), 1.
60. F. C. Frank and J. H. van der Merwe, *Proc. Roy. Soc.* (*London*) **A198** (1949), 205, 216.
61. D. H. Bradhurst and J. S. L. Leach, *J. Electrochem. Soc.* **110** (1963), 1289.
62. D. A. Vermilyea, *ibid.* **110** (1963), 345.
63. J. A. Davies, B. Domeij, J. P. S. Pringle, and F. Brown, *ibid.* **112** (1965), 675.
64. A. T. Fromhold, Jr., *Surface Sci.* **29** (1972), 396.
65. B. Borie, C. J. Sparks, Jr., and J. V. Cathcart, *Acta Met.* **10** (1962), 691.
66. D. H. Bradhurst and J. S. L. Leach, *Trans. Brit. Ceram. Soc.* **62** (1963), 793.
67. J. S. L. Leach and P. Neufeld, *Proc. Brit. Ceram. Soc.*, No. 6 (1966), 49; *Corrosion Sci.* **9** (1969), 225.
68. J. L. Ord and E. M. Lushiku, *J. Electrochem. Soc.* **126** (1979), 1374.
69. D. J. DeSmet and J. L. Ord, *ibid.* **130** (1983), 280.
70. C. G. Matthews, J. L. Ord, and W. P. Wang, *ibid.,* 285.

8

CONTINUED GROWTH OF THREE-DIMENSIONAL OXIDE

8.1. EXPERIMENTAL TECHNIQUES

The field of low-temperature oxidation has been experimentally handicapped by a lack of extensive data on simple systems. Several reasons account for this paucity of data: the difficulty of measurements, poor reproducibility of metal samples, little control of impurities (especially water vapor), and incomplete characterization of the original surface and product oxide. Fortunately, this situation is changing as semiconductor techniques and surface analytical methods are applied to the problem. The result is availability of data over extensive temperature and pressure ranges.

It is useful in a discussion of the mechanisms and theories of gas-phase oxidation at low temperatures to review the experimental techniques available. Surface analytical techniques are reviewed in Chapter 5. Methods for growing thin oxides and instruments suitable for measuring their rate of growth will be briefly covered here. Pertinent literature references provide a guide to details of the techniques.

8.1.1. Growth of Thin Oxides

The methods included in this section are suitable for growing thick or thin oxides. However, oxides less than 100 Å in thickness generally form during low-temperature oxidation, so the emphasis will be on these processes. In gas-phase oxidation at low temperatures, thermal energy is insufficient to account for continued reaction. Instead, a spontaneous electric field builds up across the oxide to provide the driving force for continued reaction. Thus, the mechanism of oxide growth at low temperatures is quite similar to anodization [1] as pointed out in Chapter 7. A major variable in the process can be the presence of water as discussed below.

Anodization [2] of a metal involves contacting the oxide through an ionic conductor such as a liquid or solid electrolyte or a gaseous plasma. For instance, a porous metallic contact separated from the metal by an oxide film has been used. The external circuit biases the metal with respect to a counterelectrode and obviates the need for electronic transport through the oxide. Thus, where ion movement in the oxide is rate limiting, anodization and thermal oxidation in the presence of a spontaneous field are similar.

Formation of thin oxide films on metals immersed in a liquid electrolyte can occur as a result of factors other than applied field. Metal can be spontaneously attacked by the dissolved oxygen in the electrolyte or, if the metal is reactive enough, by water itself. Metals more oxidizing than hydrogen in the electromotive series are capable of the latter reaction. When a counterelectrode of a different metal is connected by an electronic conductor to the metal under test, a spontaneous voltage equal to the work function difference between the metals is set up. This voltage leads to spontaneous anodization of the test metal if the counterelectrode has a larger work function than the test metal. Photoanodization of a semiconductor is a variation of this reaction [3].

A caution must be observed in the interpretation of liquid-phase anodization. It is possible for metal ions to dissolve in the electrolyte leading to a competition between oxide formation and dissolution.

Oxide thickness control based on an equilibrium between two competing processes can be achieved using gas-phase anodization [4]. An oxygen plasma at pressures of approximately 1 Torr provides both the oxidizing species and one electrical contact to the metal. An electric field is applied between the anodizing metal and the counterelectrode through an external circuit. One advantage of this method is that the anodizing metal is held at a temperature near 25°C. Simultaneous anodization and sputtering take place during exposure of the metal to the plasma. Adjustment of the relative rates of the two processes leads to an equilibrium oxide thickness in the range of 10–20 Å for Nb and Pb [5]. Sputtering of extraneous parts of the vacuum system can lead to contamination of the oxide, but this can be avoided by careful system design. A mechanism for the transport processes within the oxide has been presented by Fromhold and Baker [6].

Solid-state anodization occurs when a voltage is impressed across an oxide sandwiched between an oxidizing metal and a counterelectrode. Porous platinum was used as a high-temperature counterelectrode by Jorgensen [7] in his study of silicon oxidation and/or anodization in oxygen. Miles [8] carried out solid-state anodization at room temperature using metals containing dissolved oxygen. Jaklevic [9] believed water was responsible for his solid-state anodization of aluminum. Anodization of Nb,

Ta, and Al through films of calcia-stabilized zirconia has also been reported [10]. Yang and O'Grady [11] used yttria-stabilized zirconia as the solid electrolyte during the electrochemical oxidation of Fe and Ni. In all cases, the application of an electric field changed the rate of oxidation. Jorgensen's interpretation of his results led to Raleigh's discussion of anodization versus thermal oxidation [12]. The source of electronic charge compensation and the definition of stopping potential are discussed by Fromhold [13]. See Chapter 2.

One caution must be observed for this type of experiment since it can be a combination of thermal oxidation and solid-state anodization. The normal internal electronic path within the oxide is placed in parallel with an external electronic path. The relative impedances of the two determine which of the two processes dominates.

Other specialized methods of forming thin oxide layers exist. Ion implantation [14] creates a buried oxide layer, while direct deposition of an oxide by vacuum evaporation or chemical vapor deposition forms a surface layer. Direct deposition has not been used extensively for thin oxides, but is known to give an abrupt interface between SiO_2 and Si [15].

8.1.2. Rate Determination

Methods often used to measure the rate of growth of thin oxide films [16a] fall into five general classes: gravimetric, manometric, optical, electrical or electrochemical, and instrumental (X-ray emission, photoelectron spectroscopy, Auger analysis, and others as discussed in Chapter 5).

Gravimetric methods depend upon the sensitivity of various microbalances [17]. Vibrating quartz crystal monitors [18,19] are sufficiently sensitive to detect submonolayer coverages, especially when a porous body multiplies the surface area available. The quartz balance is well adapted to a vacuum environment for strict control of the oxidation atmosphere. Electronic vacuum microbalances are also quite useful, but sensitivity is in the tens of monolayer range, for example, 10 μg for a 1-g load [20].

Manometric techniques are based on measuring the pressure change in a reacting gas at constant volume. Both a dynamic and static technique are available. In the dynamic approach, the pressure drop for oxygen flow across an orifice is measured to determine the volume of oxidant reacting with a metal. The static technique depends on measuring the change in pressure during reaction of a fixed volume of gas. Sequential addition of aliquots of gas to the metal offers a way to monitor the rate of reaction as a function of cumulative reaction. Measurement of volume changes at constant pressure can, of course, be substituted for measurement of pressure changes. Experimentally, the constant pressure approach is more difficult.

Ellipsometry [21,22] is an important optical technique developed for examining very thin reaction layers on a metal surface. It can detect submonolayer coverages of oxygen but is also useful in the hundreds of angstroms range. Most work to date [23] has been carried out at a fixed angle of incidence and wavelength of light. Variation of these two parameters offers a further means for increasing the amount of information gathered by ellipsometry. A caution has been expressed on the role of surface roughness [24].

The measurement of interference fringes is a classical technique for studying oxide films on metals. However, it is not useful for studying low-temperature oxidation since the films produced seldom meet the thickness requirement that $x = \lambda/4n'$ where λ is the wavelength of light and n' is the refractive index of the film. The first order interference color for oxides on common metals begins to show up at a thickness above 500 Å (white light). Radiation in the 50-Å wavelength range would be needed to follow the growth of 10-Å-thick oxide films.

Reflectance spectroscopy has also been applied to characterization of thin oxide films on metals [25]. Simultaneously, identification based on absorption bands and thickness based on interference fringes have been applied to oxidized chromium [26].

There are several electrical techniques which have been applied to oxidation. Two of them are based on the resistance change of thin metal films exposed to oxygen and are discussed in Chapter 5. A third, cyclic voltammetry, is often more qualitative than quantitative. It depends on the measurement of current–voltage curves in an electrochemical cell and is used primarily in corrosion studies [27]. In this method, anodic and cathodic biases are alternately applied to a metal which forms one electrode of the cell. The growth of a protective layer is indicated by a decreasing current. Breakdown or dissolution of the oxide results in a sudden current increase, as does gas generation. A related coulometric technique is based on the quantity of electricity needed to reduce an oxide film to the parent metal [28]. Film thickness is calculated using Faraday's law.

8.1.3. Characterization of Oxide

The structure and composition of the thin oxides formed during low-temperature oxidation have been difficult to characterize in the past, but can now be analyzed using modern surface techniques. Many of these, for instance RHEED, are outlined in Chapter 5. Others [29] may be adopted as needed. For instance, Mössbauer spectroscopy has been used to profile the film formed during oxidation of iron [30] at 180°C. Scanning high-energy electron diffraction (SHEED) has been applied to the low-temperature oxidation of aluminum [31].

Exoelectron emission (see Chapter 6) and electron tunneling, in particular tunneling spectroscopy [32], have contributed to a knowledge of the electronic structure of the oxide. Tunneling spectroscopy has also been used to examine water infusion into a tunnel junction [33].

Structure of anodic oxide films has been examined using transmission electron microscopy (TEM). Dark field TEM is useful for measuring crystallite size which can be compared with determinations made using techniques such as RHEED. Microvoids, 20–40 Å in diameter, have been observed in Al_2O_3 films using TEM [34]. Similarly, 10-Å pores in SiO_2 films have been detected. See Chapter 9.

The Al/Al_2O_3 interface has been examined by high-resolution electron microscopy [35]. It was shown to have both long and short-range roughness. The former originated in the coarseness of the original aluminum surface while the latter represented ledges on the aluminum at the Al/Al_2O_3 interface. Ledge height generally did not exceed two interplanar spacings.

Optical spectroscopy provides a convenient method for examining film structure. Reflectance spectroscopy of oxide films has been analyzed by Swallow and Allen [25,26]. Infrared studies [36] offer a nondestructive technique which complements light scattering and electron spectroscopic techniques. Raman spectroscopy based on a laser light source [37] provides both chemical and structural data on films too thin for X-ray diffraction [38]. The role of surface roughness in surface-enhanced Raman spectroscopy (SERS) is not critical, and smooth metal samples may be employed [39].

Thus, the morphology, chemical composition, and electronic states in the oxide can be at least partially characterized. It is critical to the success of any study of low-temperature oxidation that the kinetic results be related to the structure of the oxide being formed.

8.2. RESULTS AND CONCLUSIONS

It is most helpful in verifying a particular model of low-temperature oxidation if several measurement techniques are applied simultaneously during the course of a particular experiment. Studies of this type are fortunately becoming available as illustrated by Figure 5.10 for iron [40] and Figures 5.5 and 7.4 for aluminum [41,42]. However, even in those cases, more extensive variation with temperature and oxygen pressure is needed to allow trial fits of data to theories of low-temperature oxidation. In the following discussion, it will be useful to summarize those metals which have been studied to a limited extent and treat in detail only those with sufficient data to allow analysis according to a particular oxidation model.

8.2.1. Role of Water

Oxidation of metals at low temperatures in the presence of moisture commands great interest from a utilitarian standpoint, but suffers from a lack of theoretical foundations. The presence of water in the oxidant and/or the oxide complicates the interpretation of low-temperature oxidation. Surface-adsorbed hydroxyls and water molecules change the measured value of surface potential as pointed out in Chapter 6. Instead of aiding ion movement, the measured potential should retard it, although we know from experiments that oxide growth continues. Fehlner and Mott [43] suggest that surface water dipoles may prevent detection of an internal field which promotes ion movement and they propose that the Cabrera–Mott theory is still applicable. On the other hand, the gel-like structure of a metal oxyhydroxide may not support charge separation which normally accounts for the field-driven process. Instead, ion movement may take place under the influence of a concentration gradient, provided the activation energy for ion movement is low.

Layers of water can be adsorbed on a metal or oxide surface in high relative humidities. These can promote lateral ion movement which sets up localized electrochemical cells due to inhomogeneities in the underlying metal. Such cells promote pitting corrosion. In addition, incorporation of water as hydroxyls in an oxide can change the mode of ion movement since a metal oxyhydroxide rather than an oxide may form. The resulting gel-like structure can strengthen the weak ionic bonding of network-modifying oxides as pointed out by Revesz and Kruger [44]. The relative stability of such metal oxyhydroxides may be estimated by comparing the free energy changes for the solvation of ions as collected in Table 8.1 taken from Gomer and Tryson [45]. The larger change in free energy per ionic charge indicates greater stability. The oxide structure so formed should slow the rate of oxidation as found in the case of copper [46]. On the other hand, water can weaken a strongly bonded, network-forming oxide as pointed out by Revesz and Fehlner [47] for silica where the rate of wet oxidation exceeds that of dry.

Okamoto and Shibata [48] suggest that the presence of water in the oxide on stainless steel may play an important role in stability of its surface against corrosion.

8.2.2. Experimental Results for Wet and Dry Conditions

A worldwide body of empirical data on corrosion at ambient temperatures has been collected as evidenced by the books of Ailor [49] and Uhlig [50]. Interpretation is generally limited to log–log presentations of data as in the work of Pourbaix [51]. He presents steel corrosion in terms of a linear bilogarithmic law:

$$\log p = A_1 + B_1 \log t \tag{1}$$

TABLE 8.1. SINGLE ION SOLVATION FREE ENERGY CHANGES [45]

Ion	$-\Delta G_s^0$ (eV)	$-\Delta G_s^0$/charge (eV)
H^+	10.98	10.98
Li^+	5.12	5.12
Na^+	3.93	3.93
K^+	3.17	3.17
Be^{2+}	24.41	12.21
Mg^{2+}	19.17	9.59
Ca^{2+}	15.69	7.85
Sr^{2+}	14.26	7.13
Ba^{2+}	13.05	6.53
Al^{3+}	46.98	15.66
Ga^{3+}	47.27	15.76
Tl^{3+}	41.50	13.83
Ce^{3+}	34.10	11.37
Fe^{3+}	44.34	14.78
Cr^{3+}	45.26	15.09
Fe^{2+}	19.13	9.57
Cu^{2+}	21.04	10.52
Cu^+	5.63	5.63
Ag^+	4.64	4.64

where p is corrosion penetration (μm) and t is time in years. Values of constants A_1 and B_1 vary with the nature of the metal and differing atmospheric conditions. Impurities, such as SO_2, are quite important in determining penetration depths over the course of several years.

When data fitting Equation 1 produces a straight line, it can also be expressed as

$$W_m = Kt^n \tag{2}$$

where W_m is mass loss (g cm^{-2}), t is exposure time (years), and K, n are constants [52]. Thus,

$$A_1 = \log K \quad \text{and} \quad B_1 = n$$

when p is converted to mass, that is, $\rho' p \times 10^{-4} = W_m$, where ρ' is oxide density (g cm^{-3}). If $n = \frac{1}{2}$, corrosion follows a parabolic rate law.

Intense interest has focused on corrosion of metallization in electronic circuitry. An early study by Campbell and Thomas [46] compares the rates of copper oxidation from 100–256°C in dry and moist oxygen. The presence of moisture was found to slow oxidation, an effect opposite to that found in silicon oxidation. See Chapter 9. Dry oxidation of copper alloys, nickel, and stainless steel were also treated in this work. A comparison of copper oxidation in clean and laboratory air, at 50–150°C, has been presented by Pinnel et al. [53]. Photoassisted corrosion of copper has also been reported [3].

Indoor atmospheric corrosion of copper, silver, nickel, cobalt, and iron at room temperature is documented by Rice et al. [54]. They emphasize the role of water and pollutants in the air such as SO_2, H_2S, NO_2, Cl_2, HCl, NH_3, and O_3. A comparison of indoor and outdoor corrosion rates for the same metals has also been published [55]. The role of relative humidity is emphasized. Available studies include detailed treatments of copper and silver [56], iron [57], cobalt [58], and nickel [59] corrosion. The latter two studies show that cobalt and nickel follow linear kinetics with exponential dependence on relative humidity. Complex oxyhydroxides form in the case of cobalt. Contact resistance of nickel in a laboratory environment has been evaluated by Pinnel et al. [60]. Cobalt oxidation in air from 25–467°C has been measured by Tompkins and Augis [61].

The oxidation of aluminum under dry conditions of 22–527°C and 10^{-9}–10^{-2} Torr oxygen was carried out by Hayden et al. [62]. They found that, after initial chemisorption and oxide nucleation, logarithmic kinetics dominated the 227–327°C range. A square-root pressure dependence was found for oxide thicknesses of 5–20 Å. Higher temperatures of 337–527°C produced parabolic kinetics with an activation energy of 1.17 ± 0.05 eV. Film thickness was 15–45 Å.

The addition of water to oxygen and their combined effect on aluminum oxidation was investigated by Grimblot and Eldridge [63]. The presence of water increased the rate of oxide growth and led to a nonzero extinction coefficient in the oxide as determined by ellipsometer measurements. This was attributed to excess Al in the oxide. These results may be compared with results for dry oxidation of aluminum [64] discussed below.

Hunt and Ritchie [65] carried out resistance marker experiments on Al_2O_3 in dry oxygen. A porous platinum electrode was deposited onto the oxide surface to serve as a counterelectrode to the aluminum. Application of a negative bias to the platinum electrode increased the rate of oxide growth; a positive bias decreased it. For the 230–400°C range, they concluded that oxide growth occurred by metal transport. The oxide was taken to be *n*-type from the sign of the Seebeck coefficient.

A concise study of the dry oxidation of indium at room temperature has been carried out by Eldridge et al. [66]. Figure 8.1 shows both the

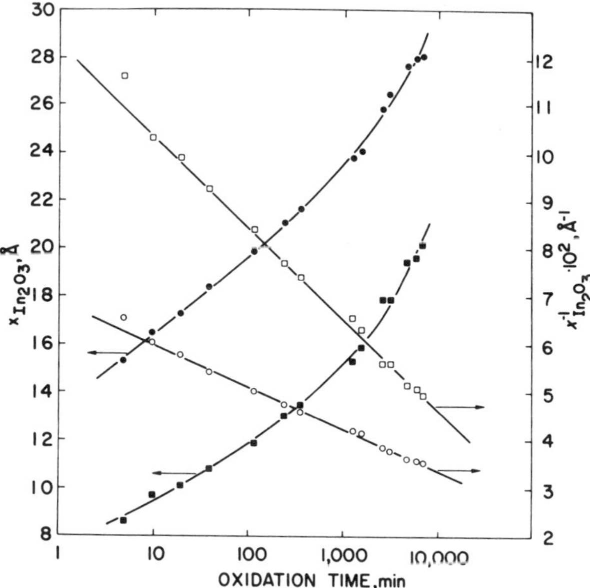

Figure 8.1. Comparison of direct and inverse logarithmic kinetics for the growth of In_2O_3 on an indium film at room temperature and 1 atm oxygen: □■ calculated taking the absorption coefficient equal to zero; ○● calculated taking the refractive index equal to 2.10 based on ellipsometer measurements. After Eldridge et al. [66]. Published by permission of the copyright holder, Elsevier Sequoia.

direct and inverse logarithmic plots of oxide thickness versus time. It appears that inverse logarithmic kinetics give the better fit. However, the interpretation of the ellipsometric results for indium oxidation requires that a choice be made between a decrease in either the real or complex part of the refractive index of In_2O_3 as oxide thickness increases. It seems more realistic to conclude that the imaginary part, that is, the absorption coefficient, decreases with increasing thickness. This implies that the metal–oxide interface is doped with excess indium and thus is an *n*-type semiconductor.

The dry oxidation of zinc at 175–402°C has been reported by Hauffe [67]. He fits the data to a parabolic rate law in contrast to the results of Vernon and coworkers [68] who fit their 25–325°C results to a logarithmic law. Zinc oxidation in air at 21°C was studied by Hammer and Shemenski [69]. The effect of an electric field on zinc oxidation was examined by Jorgensen [70] and Anderson and Ritchie [71]. The latter claim to have detected a transition from logarithmic to parabolic kinetics using this technique.

Beta-phase plutonium oxidation under dry conditions [72] initially conforms to a linear rate law in the range 120–206°C. This stage of oxidation depends on the presence of compact black α-Pu_2O_3 and has an Arrhenius activation energy of 0.67 ± 0.08 eV. The black oxide blisters as growth continues, producing a loose brownish–green oxide containing mainly PuO_2. In this latter stage, growth accelerates with time. A minimum in rate is found at ~190°C. It has been reported that the linear stage is preceded by a parabolic stage of short duration. Moisture accelerates the rate of oxidation. The combination of oxygen and water appears more effective than nitrogen and water in speeding reaction, although there has been some disagreement on this point. Linear oxidation of uranium has also been reported [73].

The reaction of bismuth with purified oxygen was found to follow a parabolic rate law between 175 and 250°C [74]. Tahboub et al. [75] agree with the parabolic kinetics, although they report an abrupt decrease in rate after 10 to 100 min. Moisture increases the rate constant without changing the activation energy. Joyner et al. [76] found during PES studies that X-rays can affect bismuth oxidation.

The oxidation of molybdenum films in dry air (70–100°C) has been interpreted in terms of dual logarithmic kinetics by Lee and Doo [77]. See Figure 8.2. In moist air, a linear rate law was found due to formation of a nonprotective, porous corrosion film.

Titanium oxidation in oxygen at 25 to 400°C was studied by Smith [78]. The data fit a logarithmic growth law for films less than 100 Å thick. The rate depended on the square root of oxygen pressure. Oxide dissolution into the metal complicated interpretations of the kinetics at higher temperatures. Gold film marker experiments were also complicated by interfering reactions between gold and titanium.

Aging of tellurium films in moist air has led to a model in which film oxidation is a function of temperature and relative humidity [79]. An activated process was found for the reaction of tellurium with dry oxygen [80]. Detectable oxide could be found after 30 min in 800-Torr oxygen only if the temperature were above 60°C. Electron irradiation caused removal of the oxide.

The oxidation of magnesium at room temperature and either 2×10^{-7} or 2.5 Torr produced a logarithmic rate [81]. Potassium oxidation followed linear kinetics at room temperature after an exposure of 5 L [82]. However, a protective oxide layer a few molecules thick formed at −196°C.

The formation of thin (<100 Å) oxide films on silicon has been an active area of investigation for device applications [83]. One of two oxide growth methods has generally been employed: low temperatures and 1 atm oxygen or high temperatures and reduced oxygen partial pressures in nitro-

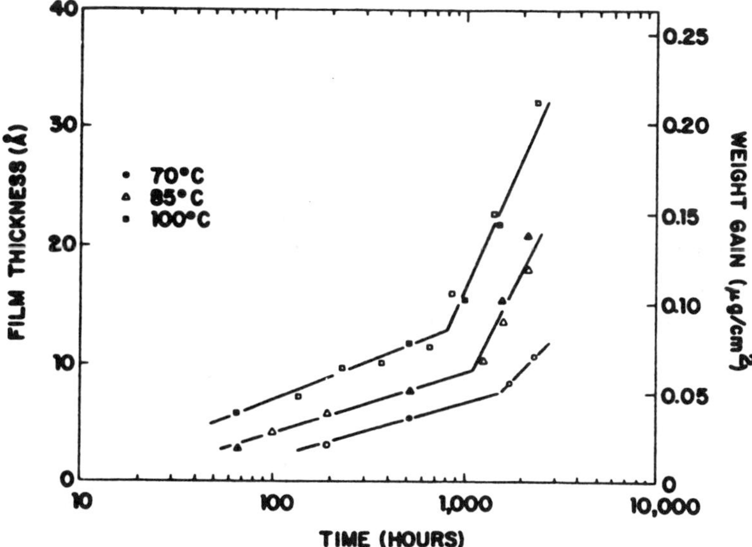

Figure 8.2. Oxidation of sputtered molybdenum films in $C_5F_{11}NO$ (open symbols) and dry air (solid symbols). Weight gain data were calculated from the oxide thickness assuming MoO_2 is formed ($\rho' = 6.47\ cm^{-3}$). After Lee and Doo [77]. Reprinted by permission of the publisher, The Electrochemical Society, Inc.

gen. The more detailed studies are treated in Section 8.2.3.2. However, a summary of other studies is appropriate at this point.

Early work on silicon has been collected and published by Fehlner [84]. Later work based on an AES study [85] identified four stages of dry oxidation: monolayer formation, intermediate oxide, formation of a three-dimensional SiO_2 film, and bulklike SiO_2. Some effect of electron beam irradiation was noted.

Oxidation of cleaved single crystal silicon in air at room temperature has been carried out by Lukeš [86]. He interpreted his ellipsometer results in terms of a logarithmic growth rate which changed to a cubic rate law after 10^7 sec. Relative humidity varied between 40 and 90%. He found a value of ~7 Å per decade for the slope of oxide thickness plotted versus the logarithm of time, in agreement with Archer [87].

Mende et al. [88] studied the air oxidation of single crystal silicon etched by hydrofluoric acid. Measurements based on XPS and nuclear activation analysis agreed, giving a logarithmic growth rate of 2 Å per decade. Raider et al. [89] point out that ellipsometric measurements tend to overestimate oxide thickness due to impurity films. Fehlner [90] also cautions that HF etching in the presence of UV irradiation can give rise to unexpected film growth.

The oxidation of amorphous silicon ($a-\mathrm{Si:H}$) in room temperature air was carried out by Ponpon and Bourdon [91] using ellipsometry. Their results closely followed those of Mende et al. [88] up to 6×10^4 sec. Results were interpreted in terms of logarithmic kinetics up to $\sim 10^5$ sec and a parabolic rate thereafter. Kelemen et al. [92] found agreement between the oxidation rate of $a-\mathrm{Si:H}$ and single crystal silicon in dry oxygen at 25°C, but at 300°C the a-Si:H oxidized faster.

Formation of thin SiO_2 has also been carried out at high temperatures and reduced pressures [93] and at intermediate temperatures with pressures greater than 1 atm [94]. The catalytic nature of Au on the initial formation of $SiO_{4/2}$ tetrahedra is also of interest [95].

8.2.3. Comprehensive Studies of Dry Oxidation

Analysis and interpretation of detailed kinetic data pose a problem. Historically, linearity of a plot on graph paper was taken to be indicative of a particular mechanism. However, it has become apparent that, for low-temperature oxidation, both logarithmic (x versus log t where x is oxide thickness, t is time) and inverse logarithmic kinetics ($1/x$ versus log t) can often fit the same set of data. One resolution of this dilemma is to examine the principles upon which a derived rate equation is based and calculate values for variables which can be checked with values from other experiments. Several descriptive models of low-temperature oxidation have been put forward as discussed in Chapters 2 and 7, but the Cabrera–Mott theory stands out as being well developed. For this reason, it is emphasized here for interpreting kinetic data from low-temperature oxidation.

An examination of the experimental data available at the present time indicates that metals which are carefully prepared to avoid impurities can, at sufficiently low temperatures, produce microcrystalline or vitreous oxides with a uniformity which matches the assumptions of the Cabrera–Mott theory. This is especially true for network-forming and intermediate oxides below several hundred degrees Celsius, but may also be true for modifying oxides at cryogenic temperatures. Thus, kinetic studies have been successfully interpreted by Fehlner [96] in terms of the Cabrera–Mott theory in the form proposed by Ghez (Equation 28 of Chapter 2). Other literature results based on numerical integration are included for comparison. The present study is restricted to dry oxidation where data are available over a sufficiently broad span of temperature.

Data points were taken from the original curves in the literature using a digitizing technique. Values of x and t for a particular T were stored on magnetic tape until needed. A computer program was used to convert the

TABLE 8.2. VALUES OF N, W, AND $|V|$ OBTAINED FROM KINETIC DATA FOR LOW-TEMPERATURE OXIDATION [96]

Group no.	Element	Cation[a] transport number	N (cm^{-2})	W (eV)	$\|V\|$ (V)	Reference
IA	Na	—	1×10^{-8}	−0.2	0	158
IB	Cu	~1	1×10^{21}	1.9	—	155, 156
IB	Cu		2×10^{7}	0.5	86	46
IIA	Be	0.75	—	—	—	159
IIIA	Al	0.4–0.6	6×10^{8}	1.6	~0.6	115, 116
IIIA	Al		2×10^{3}	~0.9	~3.0	64
IVA	Si	<0.05	1×10^{6}	1.5	2.6	84, 99, 102
IVA	Ge	—	2×10^{6}	1.1	0.2	97
IVA	Pb	—	7×10^{2}	0.4	1.5	109, 110
VB	Ta	0.24	3×10^{12}	1.6	1.5	152, 153
VIB	Cr	—	5×10^{13}	1.8	2.8	123
VIII	Fe	—	5×10^{8}	0.8	—	134
VIII	Ni	—	3×10^{13}	1.6	2.3	142

[a] From Table 4.1.

raw data into values of $1/x$ and $\log(t/x^2)$ (x in Å and t in sec). These points were then plotted and values of the slope m and intercept b determined based on a least-squares analysis. Values of correlation coefficient R were also determined. A value of 1 indicated perfect correlation.

Values of u and x_1 were calculated from m and b and plotted versus $1/T$ according to Equations 29 and 30 of Chapter 2. The slope based on Equation 29 was used to determine activation energy W while the intercept was used to calculate a value of N. The vibrational frequency ν was taken to be 10^{12} sec^{-1} while Ω was derived from the density of the relevant oxide.

The absolute value of V was determined from Equation 30 using the slope of x_1 versus $1/T$. The value of $2a$ was assumed to be approximately equal to the ionic spacing in the crystalline oxide. This assumption, as well as the one concerning Ω, are considered valid since short-range order is maintained in bulk vitreous oxides. This is expected to hold true for similar oxides formed as films during low-temperature oxidation.

The results of the study [96] in terms of values of W, N, and $|V|$ are listed in Table 8.2. The activation energy W is compared in the text with literature values determined at high temperatures, although it is not altogether clear that either the parabolic or linear rate constants are directly relevant. The Cabrera–Mott equation assumes interface control through incorporation of the mobile ion while the parabolic equation is based on control by diffusion in the bulk oxide. The linear rate constant depends on interface control, but at the reaction interface rather than the incorporation one. The comparison is, therefore, valid only if both the low- and high-temperature processes depend on the same key event such as breaking a metal–oxide or metal–metal bond.

The best comparison of W may be made with values reported for anodization [2] in cases where high resistivity oxide is formed. The similarity between anodization and low-temperature oxidation lies in the fact that both are field-driven processes. As a result, direct comparison between

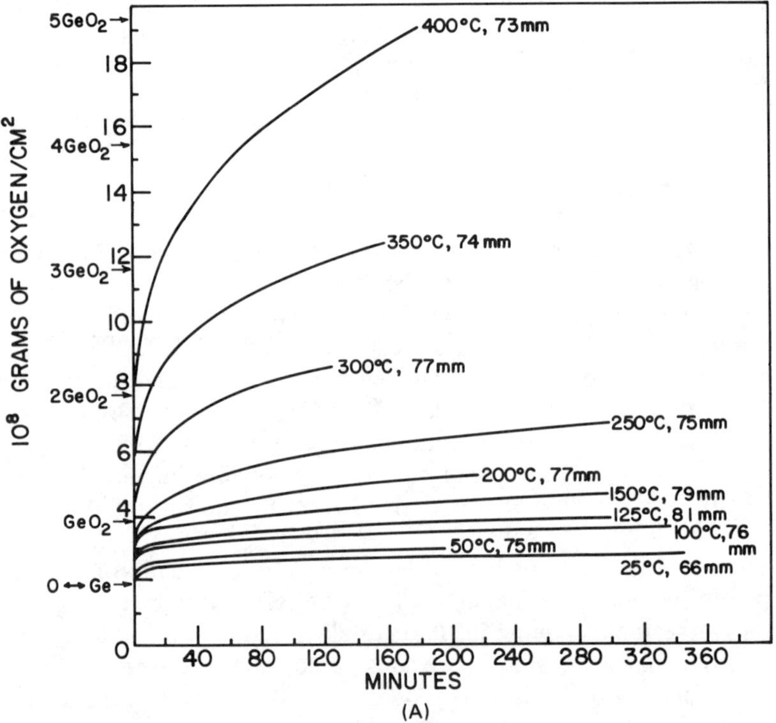

Figure 8.3. Germanium oxidation. (A) Oxide growth on germanium powder as a function of temperature. Reprinted with permission from *J. Phys. Chem.*, Vol. 64, p. 1020, J. R. Ligenza, The Initial Stages of Oxidation of Germanium, Copyright 1960, American Chemical Society. (B) Plot of log u versus $1/T$. (C) Plot of x_1 versus $1/T$. After Fehlner [96]. Reprinted by permission of the publisher, The Electrochemical Society, Inc.

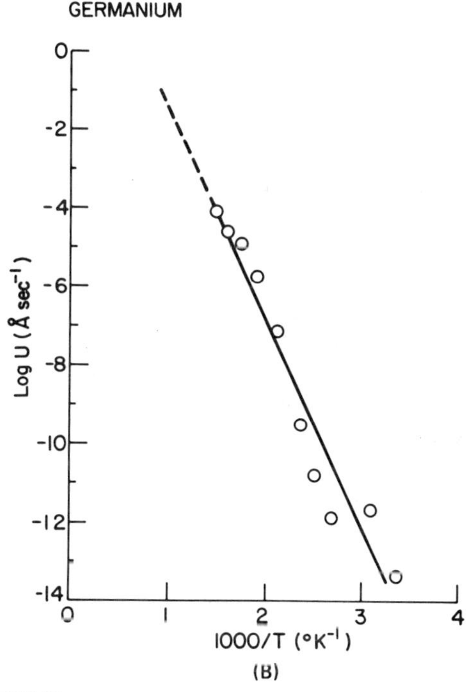

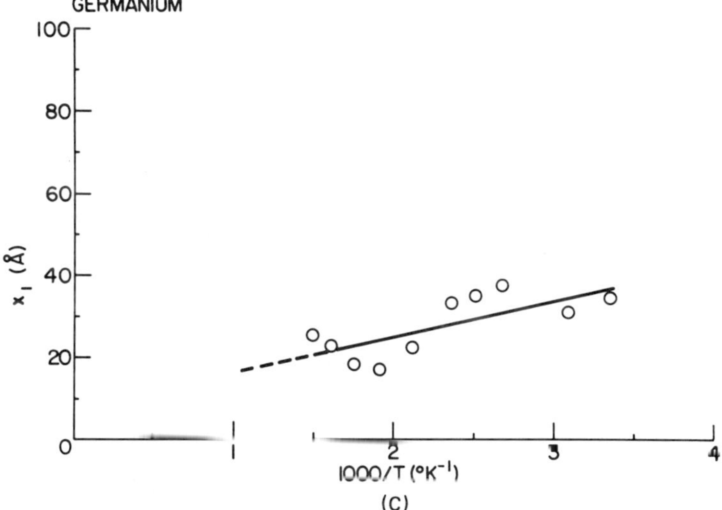

Figure 8.3. (*Continued*)

the two is possible. However, such results are scarce due to temperature restrictions and the limited ability of some metals to anodize.

8.2.3.1. Germanium. Data for the low-temperature oxidation of germanium, as shown in Figure 8.3A, were taken from the work of Ligenza [97]. He measured the rate of dry oxidation for germanium powder at 75 Torr between 25 and 400°C. The data were analyzed using the Ghez formulation of the Cabrera–Mott equation, $1/x$ versus $\log(t/x^2)$, as well as with a plot of x versus $\log t$. No choice could be made on the basis of correlation coefficients between the two kinetic expressions.

For reasons stated previously, the data were further analyzed using the Cabrera–Mott expression. Hence, values of $\log u$ were plotted versus $1/T$ as shown in Figure 8.3B. The correlation coefficient was 0.970. A value of $W = 1.1$ eV was calculated from the slope and a value of $N = 2 \times 10^6$ cm^{-2} from the intercept taking $\nu = 10^{12}$ sec^{-1} and $\Omega = 41$ (Å)^3 (hexagonal GeO_2). This value of W is difficult to compare with activation energies obtained at high temperatures since the volatility of the suboxide leads to linear kinetics [16b]. As a result, the activation energy obtained at high temperatures is not representative of ionic movement during diffusion as represented by parabolic kinetics.

It is interesting to note that the present analysis gives rise to a temperature independent activation energy for ion movement as contrasted with Ligenza's analysis where W was a function of T.

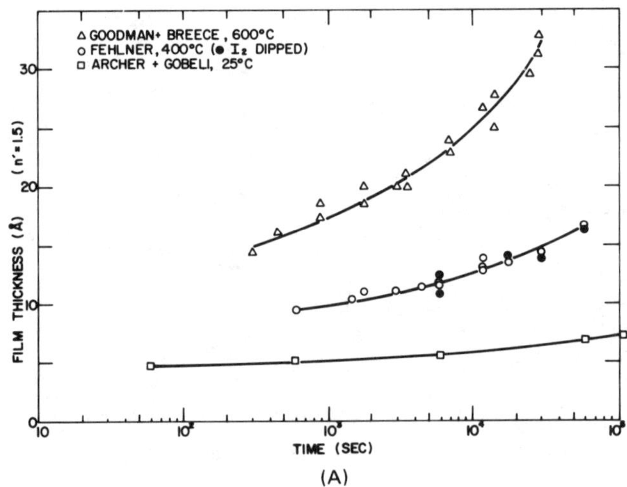

Figure 8.4. Silicon oxidation. (A) Oxide growth on single crystal silicon as a function of temperature. After Fehlner [84]. (B) Plot of $\log u$ versus $1/T$. (C) Plot of x_1 versus $1/T$. After Fehlner [96]. Reprinted by permission of the publisher, The Electrochemical Society, Inc.

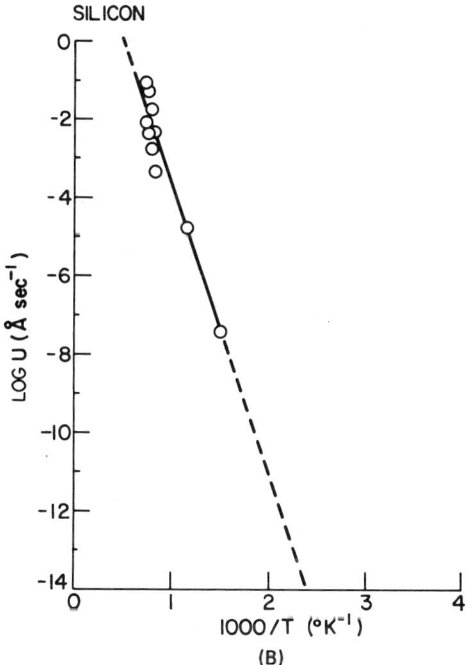

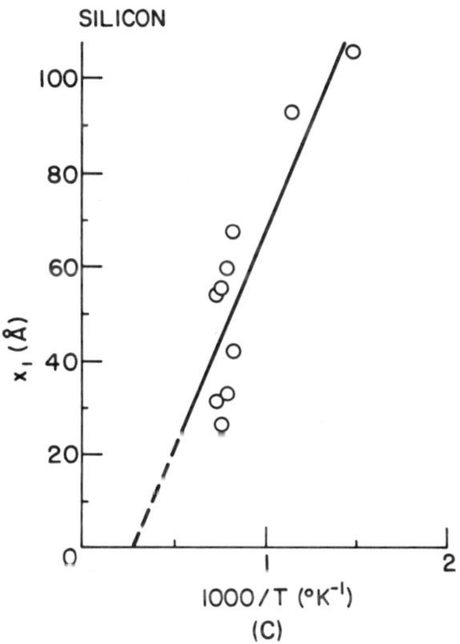

Figure 8.4. (*Continued*)

The plot of x_1 versus $1/T$, shown in Figure 8.3C, produced a correlation coefficient of 0.736. Taking $Z = 2$ for anion conduction as discussed below and $2a = 3$ Å, $|V|$ was determined to be 0.2 V. This is close to the surface potential value of −0.3 V obtained by field emission for monolayer coverage of oxygen on germanium [98].

8.2.3.2. Silicon. The 1972 paper of Fehlner [84] summarized literature data for low-temperature oxidation of single crystal silicon at 25, 400, and 600°C as shown in Figure 8.4A, but these data were insufficient to define the kinetics. Kamigaki and Itoh [99] found that silicon oxidation between 950 and 1100°C followed Cabrera–Mott kinetics provided that the oxygen partial pressure in an O_2/N_2 mixture held at 1 atm was less than 10^{-1} atm (~76 Torr). In a subsequent paper by Horiuchi et al. [100], Cabrera–Mott kinetics were found under similar conditions, even when the oxygen supply contained less than 2×10^{-26} atm water vapor. Hopper et al. [101] interpreted their detailed data in the ranges 0.2–1.6 atm oxygen and 780 to 930°C in terms of the linear-parabolic equation. No fit of their data to the Cabrera–Mott equation was found in the present work. As a result, only the data of Kamigaki and Itoh at 10^{-2} and 10^{-3} atm were analyzed along with the 1-atm data at 400°C [84] and 600°C [102].

The results of Kamigaki and Itoh throw into question the ~500°C upper temperature limit for Cabrera–Mott kinetics. The explanation for this discrepancy may lie in the nature of the diffusing species. Field-driven oxidation at lower temperature relies on ion movement [7] while oxygen molecules are taken to be the mobile species at higher temperatures [103,104,105]. See Chapter 9. At 1 atm pressure of oxygen, the temperature of transition from control by ionic to control by molecular movement is estimated to be 500°C.

Oxygen solubility in SiO_2 varies directly with pressure. Thus, a pressure reduction to 10^{-2} atm would decrease the O_2 concentration in SiO_2 a factor of 100 and it might be expected that the temperature of transition from ionic to molecular movement would rise. As a result, the change in kinetics would also occur at a higher temperature.

The Cabrera–Mott fit was slightly better in a comparison of direct and inverse log kinetics. The analysis of log u versus $1/T$ in Figure 8.4B yielded $R = 0.96$. The value of W was calculated to be 1.5 eV which may be compared with the values of 1.8 and 1.4 eV found by Lie et al. [105] for parabolic oxidation (800–900°C and 900–1000°C, respectively). They also found a value of 1.9 eV for linear oxidation over the same temperature range. The rate-limiting step in the parabolic oxidation is taken to be diffusion of a reactant through the oxide while that for the linear oxidation is taken to be reaction at the silicon–oxide interface. The 1.5 eV value for logarithmic kinetics falls between the two and may be compared with the Si–Si bond energy of 1.8 eV [106] and the O–O bond energy of 1.44 eV [107].

The value of N was calculated assuming $\Omega = 45$ (Å)^3 (vitreous silica). It was found to be 1×10^6 cm^{-2}. Voltage was determined from the slope of x_1 versus $1/T$ in Figure 8.4C where the correlation coefficient was 0.858. The value of $|V|$ was 2.6 V, taking $Z = 2$ and $2a = 3$ Å. This may be compared with the value of -0.67 V for (100) silicon or $+0.01$ V for (111) silicon [108].

8.2.3.3. Lead. A detailed study of lead oxidation has been carried out as part of the tunneling device investigation at IBM [109,110,111]. Ellipsometry was used to follow the growth rate of PbO in which anionic species are more mobile than cationic species. The oxide was grown on single crystal lead films at temperatures between -90 and 150°C. This is shown in Figure 8.5A. Comparative oxide thicknesses determined by ellipsometry, X-ray absorption, and electron tunneling all agreed within an order of magnitude. A small dependence of PbO growth rate on oxygen pressure at 25°C was also determined for the range 4×10^{-2} to 7.6×10^2 Torr.

Eldridge and Dong [109,110] had calculated values of u and x_1, but they applied them to an oxidation model based on piezoelectric effects. In the present approach, the values of u and x_1 were analyzed using the Cabrera–Mott equation.

The plot of log u versus $1/T$ (Figure 8.5B) gave a value of $W = 0.44$ eV. This may be compared with a value of 0.56 eV reported by Archbold and Grace [112] for parabolic oxidation of molten lead (453–643°C). Anderson and Tare [113] report a value of 0.48 eV for linear kinetics between 230 and 275°C. The rate was also proportional to the first power of pressure between 10^{-2} and 10^{-5} Torr. At lower temperatures, they plotted their data according to a direct logarithmic law.

The value of N was found to be 7×10^2 cm^{-2} based on $\Omega = 38.9$ (Å)^3 (tetragonal PbO). The plot of x_1 versus $1/T$ (Figure 8.5C) contained data points which were all bunched together. Since Equation 30 of Chapter 2 implies an intercept of (0,0), an estimate of $|V|$ was obtained by drawing a straight line through the origin and the center of the experimental points. A value of $|V| = 1.5$ V was estimated for anion conductivity using $2a = 4$ Å and $Z = 2$. This may be compared with the surface potential value of $+0.6$ V reported for oxygen adsorption on a lead film at 23°C and 6×10^{-3} Torr [114].

8.2.3.4. Aluminum. Dry oxidation of aluminum foil at 1 Torr has been treated by Dignam et al. [115,116] for the temperature range 250–

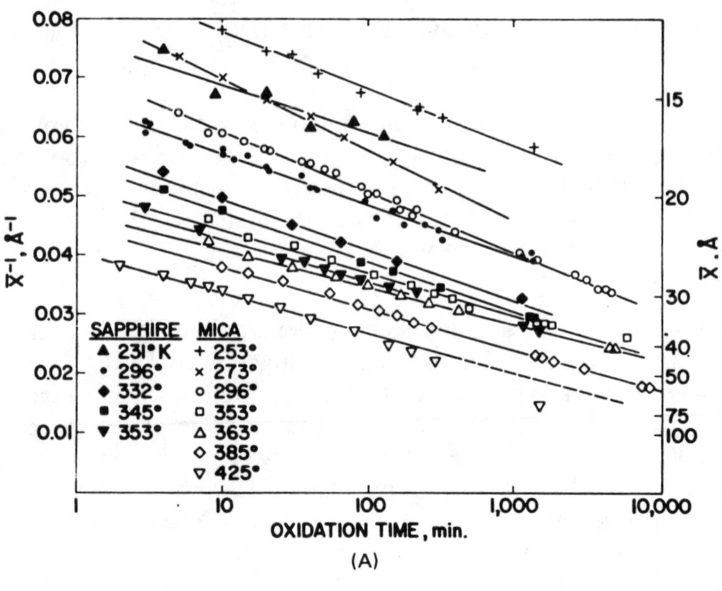

0.08
0.07
0.06
0.05
0.04
0.03
0.02
0.01
$\bar{X}^{-1}$, Å^{-1}
15
20
30
40
50
75
100
$\bar{X}$, Å
SAPPHIRE
▲ 231°K
• 296°
◆ 332°
■ 345°
▼ 353°
MICA
+ 253°
× 273°
○ 296°
□ 353°
△ 363°
◇ 385°
▽ 425°
1
10
100
1,000
10,000
OXIDATION TIME, min.

(A)

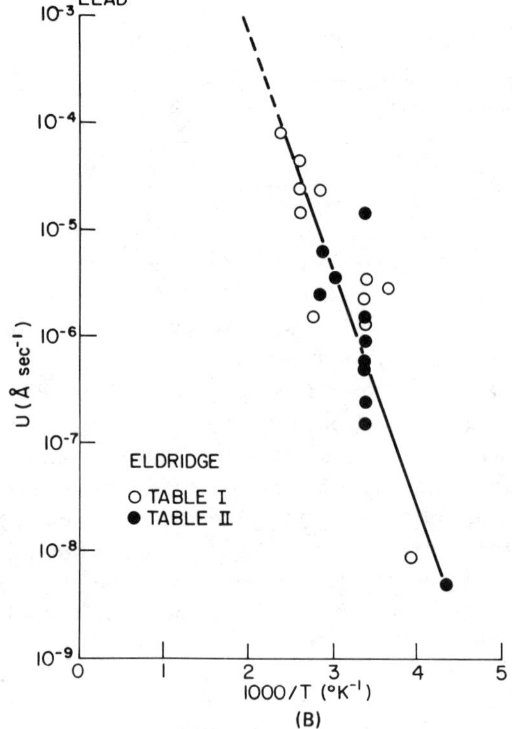

LEAD
10^{-3}
10^{-4}
10^{-5}
10^{-6}
10^{-7}
10^{-8}
10^{-9}
U (Å sec^{-1})
ELDRIDGE
○ TABLE I
● TABLE II
0
1
2
3
4
5
1000/T (°K^{-1})

(B)

Figure 8.5.

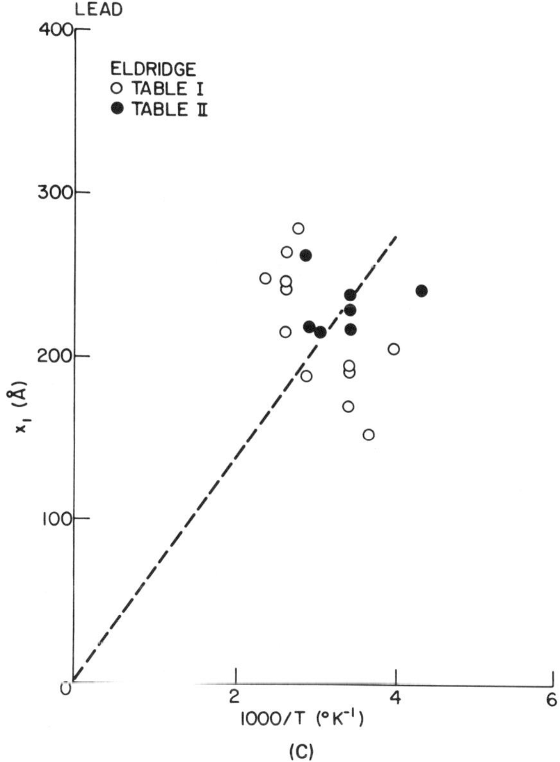

Figure 8.5. Lead oxidation. (A) Oxide growth at 1 atm O_2 on lead films as a function of temperature. After Eldridge and Dong [109,110]. The sapphire and mica were substrates for the lead films. Published by permission of the copyright holder, North Holland Physics Publishing. (B) Plot of log *u* versus $1/T$. (C) Plot of x_1 versus $1/T$. After Fehlner [96]. Reprinted by permission of the publisher, The Electrochemical Society, Inc.

505°C. A time–temperature dependent relationship was found for the transition from an "amorphous" oxide to γ-alumina crystallites as shown in Figure 8.6A. The sigmoidal kinetics characteristic of this process have been discussed in terms of oxide crystal nucleation and growth by Pryor [117], Bartlett [118], and others. The initial growth kinetics prior to crystal formation could be fitted by Dignam et al. only with an equation derived from the Cabrera–Mott theory using computer-aided procedures. Values for the constants were obtained as follows [115]:

$$2\Omega v a n_c = 2.5 \times 10^6 \text{ Å sec}^{-1}$$

$$W = 1.60 \text{ eV}$$

$$2qaV = 8.2e \text{ Å V}$$

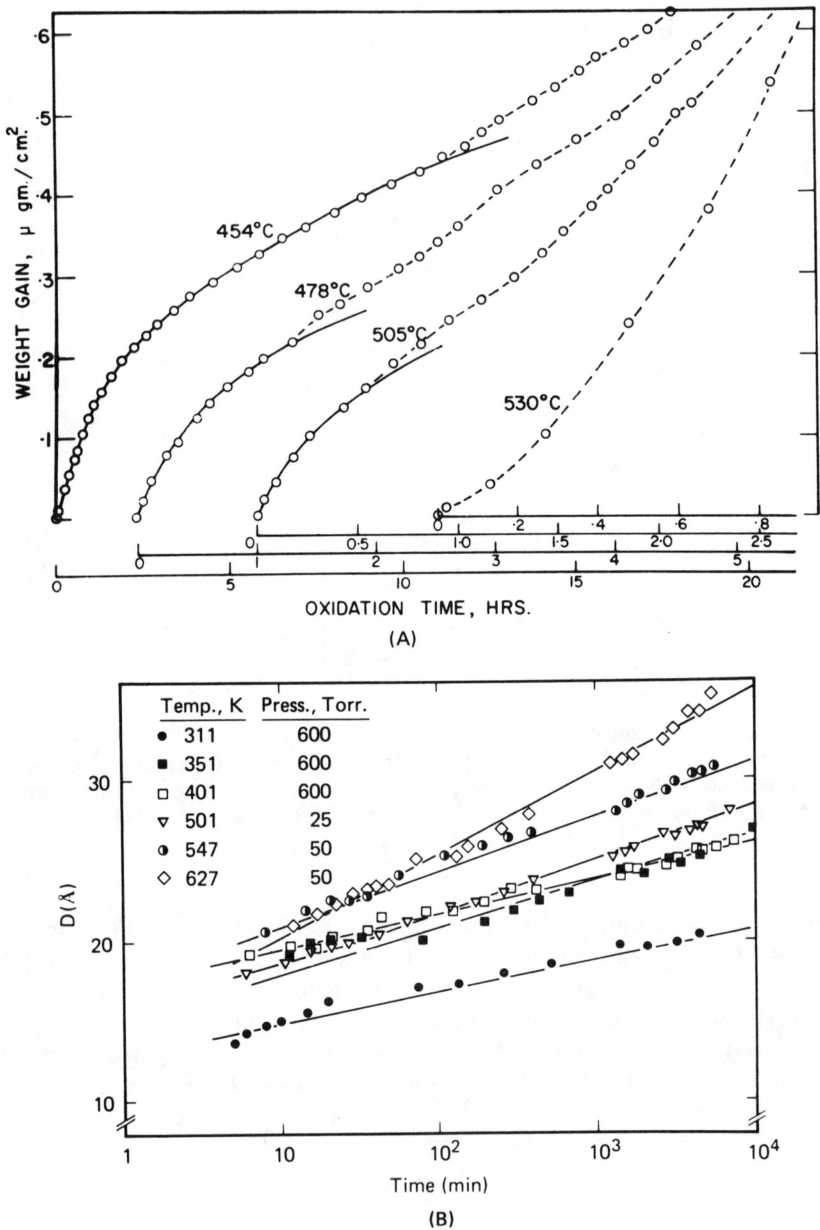
WEIGHT GAIN, μ gm./cm.²
·6
·5
·4
·3
·2
·1
454°C
478°C
505°C
530°C
OXIDATION TIME, HRS.
(A)
Temp., K
Press., Torr.
311 600
351 600
401 600
501 25
547 50
627 50
D(Å)
Time (min)
(B)

Figure 8.6.

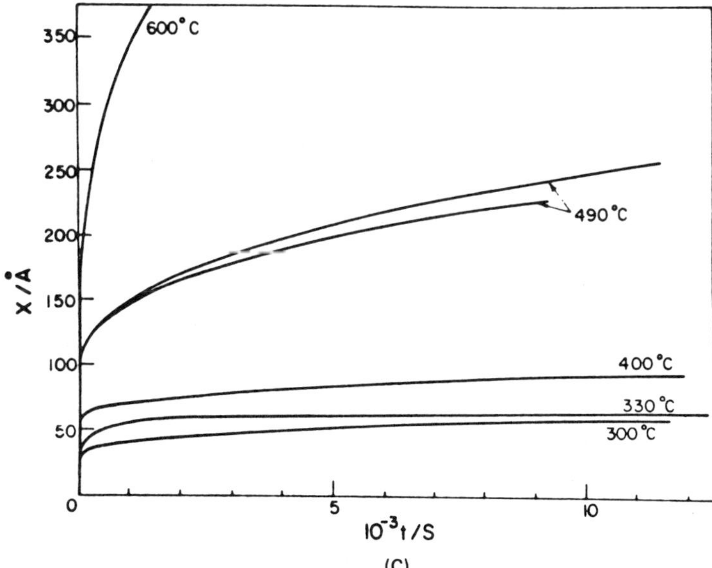

Figure 8.6. Aluminum and chromium oxidation. (A) Oxide growth on aluminum foil as a function of temperature. One μg cm^{-2} O_2 is equivalent to 53.6 Å bulk Al_2O_3. After Dignam et al. [115]. (B) Oxide thickness *D* on aluminum film as a function of temperature and pressure. After Grimblot and Eldridge [64]. (C) Oxide growth on polycrystalline chromium at 4×10^{-3} Torr oxygen as a function of temperature. After Young and Cohen [123]. All three works reprinted by permission of the publisher, The Electrochemical Society, Inc.

The value of W may be compared with 1.3 eV determined from anodization [2*a*]. In this calculation, n_c is the volume concentration of the mobile species and Ω is oxide volume per ion. For $\Omega = 45.6$ $(\text{Å})^3$, $\nu = 10^{12}$ sec^{-1}, and $2a = 5.6$ Å; $n_c = 9.8 \times 10^{15}$ cm^{-3} ($N = 2an = 5.6 \times 10^8$ cm^{-2}). Assuming $q = 3$ for Al^{3+} ion movement, $|V|$ equals 0.49 V. If $q = 2$ for anion movement, then $|V| = 0.73$ V. These values agree in general with the magnitude of values obtained through surface potential measurements. Both positive and negative surface potential values have been reported, depending upon whether polycrystalline rods [119] (+0.15 V) or films [41,120] (−0.3 to −0.7 V) were examined. See Table 6.1.

The rate of aluminum oxidation at low temperatures was reported to be dependent on the square root of oxygen pressure [121].

Grimblot and Eldridge [64] report on a study of aluminum film oxidation for conditions of 27–427°C and 25–600-Torr oxygen as shown in Figure 8.6B. They interpret their results in terms of the Cabrera–Mott equation. However, different values of the derived constants were found: W was

0.8–1.0 eV, $|V|$ was ~3 V, and N can be calculated from their results to have been 2.2×10^3 cm^{-2}. The lack of agreement between their values and those of Dignam et al. [115,116] reflects the fact that Grimblot and Eldridge found larger values of oxide thickness at lower temperatures, perhaps due to the use of different O_2 pressures. The latter authors also report that the initial oxide formed is heavily doped with aluminum and is therefore not characteristic of Al_2O_3. Eberhardt and Kunz [122] report a similar nonstoichiometry during initial oxide formation.

8.2.3.5. ***Chromium.*** Oxidation of chromium in dry oxygen at 300–600°C has been reported by Young and Cohen [123] for a pressure of 4.0×10^{-3} Torr. This is shown in Figure 8.6C. A transition from Cabrera–Mott to parabolic kinetics at ~450°C was found to be a function of both time and temperature. It is interesting to note that atmospheric oxidation of chromium between 270 and 490°C was interpreted in terms of direct logarithmic kinetics [124]. The transition could be correlated with a change in oxide structure from a flat, fine-grained, almost vitreous Cr_2O_3 to a strongly preferred fiber texture containing ridges and nodules. Since the activation energy for oxide growth in both regions was less than that for self-diffusion of Cr in Cr_2O_3 [125], the rate-controlling process appeared to be transport via preferred pathways. Numerical integration of the Cabrera–Mott expression [126] led to two equations, one based on interstitial cations and the other on cation vacancies. The former depended on N, the number of surface metal atoms in sites suitable for movement into the oxide, while the latter depended on the number of adsorbed anions on the oxide. The major difference between the two kinetic expressions lies in a preexponential factor containing $1/x$. The cation mechanism follows Equation 22 of Chapter 2, while the vacancy mechanism adds the term $1/x$ to the preexponential of the same equation.

If cation interstitials were assumed, then $N = 5 \times 10^{13}$ cm^{-2} and $W = (W_i + U) = 1.8$ eV, based on $\Omega = 48.1$ (Å)^3. The activation energy for ion movement W was the sum of W_i, the heat of solution of an interstitial cation, and U, the energy barrier to movement within the oxide.

Values of $|V| = 2.8$ V for cation interstitials and $|V| = 2.4$ V for vacancies were derived from the data. These may be compared with the values of -2.1 to -2.4 V obtained from surface potential measurements [127,128] at 25°C.

The vacancy mechanism gave an activation energy of $Q' \cong 1.9$ eV for small x, where Q' is the activation barrier to ion movement at the oxide–gas interface. The above values of activation energy may be compared with the high-temperature values for Cr_2O_3 as obtained by Gulbransen and Andrew [129] at 76 Torr. For the first 5000 Å of oxide growth, W was found to be 1.63 eV based on parabolic kinetics.

8.2.3.6. Iron. The oxidation of iron at low temperatures has been reviewed by Cohen [130] from the standpoint of both electrochemical and thermal oxidation. Epitaxial oxide of shifting composition, Fe_3O_4 to α-Fe_2O_3,was reported [130,131,132] for dry thermal growth below 200°C although there has been some discussion of the possible existence of FeO [40,132,133]. Variation in rate with crystal face for the range 24–200°C is shown in Figure 6.5 for dry oxidation [134]. The rate of reaction was plotted according to a direct logarithmic law, but no attempt was made by the authors to interpret the mechanism of reaction. The data are analyzed in more detail below. In the ranges 25–150°C and 5×10^{-8} to 2×10^{-6} Torr oxygen, the oxidation rate was directly proportional to pressure.

A transition from logarithmic to parabolic kinetics [134,135] occurred at ~200°C. The rate changed with time due to pressure dependent nucleation of α-Fe_2O_3. Initially, the rate increased directly as the square root of oxygen pressure, but it became independent of pressure ($P > 10^{-5}$ Torr) after growth of a continuous layer of α-Fe_2O_3. A difference in oxidation behavior with source of iron was ascribed to grain size and crystal orientation. The parabolic rate had an activation energy of 1.39 eV (200–400°C). Needham et al. [136] interpreted this same oxidation range in terms of dual-logarithmic growth. They found larger values of oxide thickness than reported above. The role of impurities such as carbon plus the morphology of crystal grains may account for the difference in values of oxide thickness found by the two sets of authors.

The nature of air-formed oxide films on iron has been investigated by electrometric and electron diffraction studies [137] as well as by photoelectron spectroscopy. Turner et al. [138] report that the preadsorption of water can interfere with the adsorption of oxygen on iron. Kruger and Yolken [139] found that moisture decreased the rate of iron oxidation.

Bhavani and Vaidyan [140] examined the influence of an electric field on the room temperature oxidation of iron films in air. Relative humidity varied between 70 and 90%. No counterelectrode was deposited on the iron oxide. Instead, capacitive coupling to ground served to induce a voltage, calculated to be ~0.7 V, across the oxide. The oxidation rate was enhanced by making the iron positive with respect to ground. A negative potential retarded the rate. Stress in the iron films caused discontinuities in the resistance measurements used to monitor oxide thickness. The authors suggest that growth of ferric oxide is due to oxygen ion diffusion.

A dynamic stress measurement was carried out during the oxidation of iron films at 250–350°C and 5×10^{-4} Torr dry oxygen [141]. Only

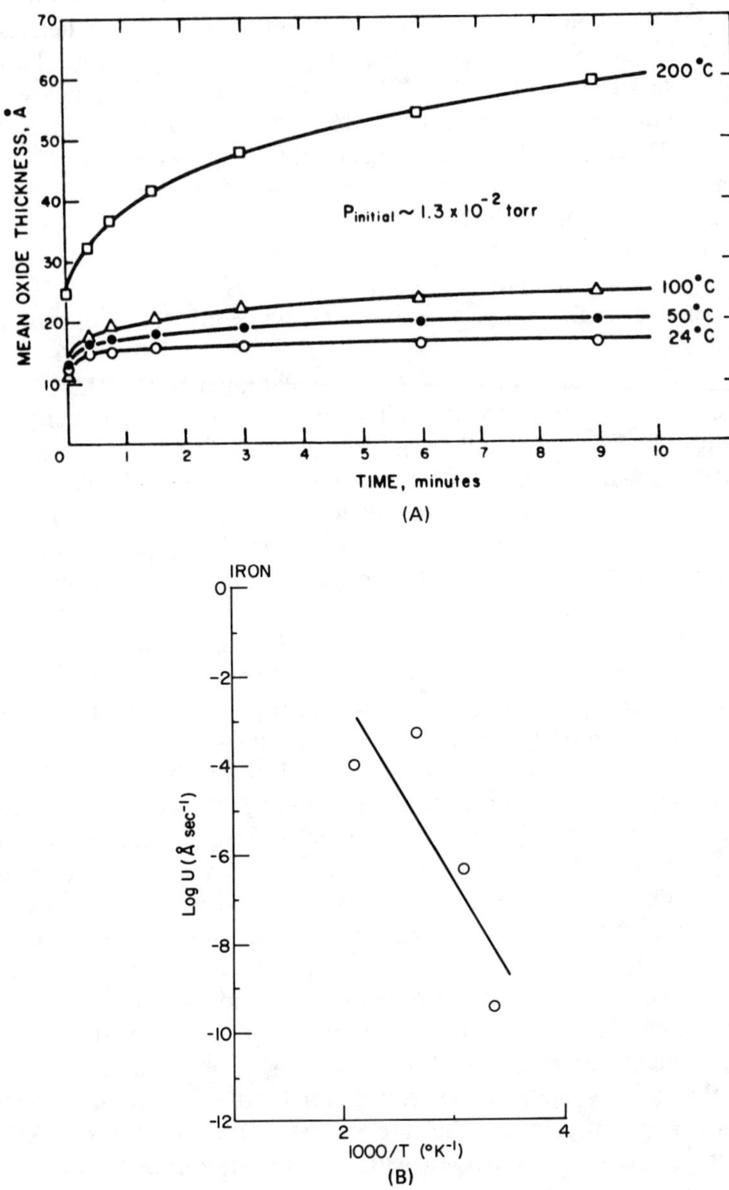

Figure 8.7. Iron oxidation. (A) Oxide growth on polycrystalline iron as a function of temperature. After Graham et al. [134]. Reprinted by permission of the publisher, The Electrochemical Society, Inc. (B) Plot of log u versus $1/T$. (C) Plot of x_1 versus $1/T$.

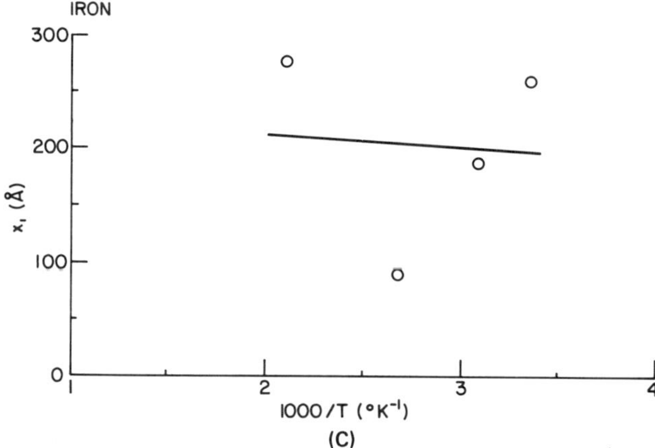

Figure 8.7. *(Continued)*

Fe_3O_4 was detected. It had a tensile stress of 10^{10} dyne cm^{-2} for a 500 Å thick film at 300°C.

Results for polycrystalline iron, for which there is the most data [134] (Figure 8.7A), were plotted according to inverse and direct logarithmic laws. In both cases, correlation coefficients were greater than 0.940, except for the direct log plot at 200°C which had a value of 0.887. Hence, the indirect log plot appeared to be a slightly better fit.

Values of the constants were found using $Z = 2$, $2a = 8.3$ Å, and $\Omega = 25$ (Å)^3 (cubic Fe_3O_4) and the plot of log u versus $1/T$ (Figure 8.7B). The activation energy was calculated to be 0.8 eV. This compares with a value of 1.39 eV as discussed above for parabolic oxidation. The polycrystalline nature of the iron film and the polycrystalline oxide growing epitaxially on it may explain the difference in results. On the other hand, the amount of data is limited, making the results imprecise.

Values of N equal to 5×10^8 and $|V|$ equal to 0.1 V were also found, the latter from the plot of x_1 versus $1/T$ (Figure 8.7C). Unfortunately, the correlation coefficient for this plot was only 0.07 so that the value of voltage is unreliable. On the other hand, values of N and W are derived from the log u versus $1/T$ plot which had a correlation coefficient of 0.83.

8.2.3.7. ***Nickel.*** The dry oxidation of polycrystalline nickel in the range 24–450°C has been studied by Graham and Cohen [142]. Results at temperatures up to 400°C and pressures of 5×10^{-3} to 6×10^{-1} Torr are analyzed below. Impurities such as C, Si, and S impeded oxidation. They were eliminated using ultrahigh vacuum techniques combined with

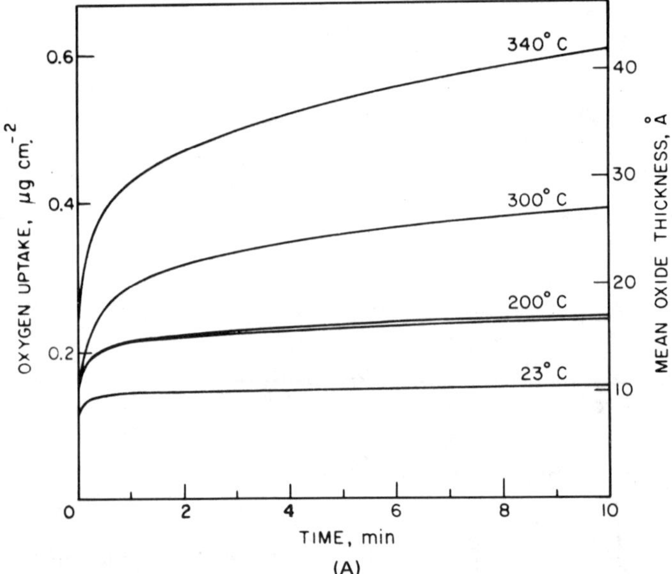

Figure 8.8. Nickel oxidation. (A) Oxide growth on zone refined, polycrystalline nickel sheet at 5×10^{-3} Torr oxygen as a function of temperature. Oxide thickness was calculated from oxygen uptake assuming NiO ($\rho' = 6.75$ g cm^{-3}, roughness factor of 1). After Graham and Cohen [142]. Reprinted by permission of the publisher, The Electrochemical Society, Inc. (B) Plot of log u versus $1/T$. (C) Plot of x_1 versus $1/T$.

chemical treatments [143,144]. Initial fast growth of two to three layers of oxide by place exchange was followed by direct logarithmic kinetics for the range 40–300°C [145,146].

Parabolic growth [142,143,147] was found for nickel oxide thicknesses greater than 30–40 Å. An activation energy of approximately 1.60 eV was determined for this parabolic rate (300–700°C), a value more characteristic of growth via paths of easy-ion movement when compared with the lattice diffusion value of ≥2.25 eV. A $P^{1/6}$ pressure dependence was found at 450°C, consistent with the formation and movement of doubly charged cation vacancies [142]. Grain boundary diffusion was an important factor in the parabolic range and variations in oxide grain size may account for differing rate laws reported by various authors for nickel oxidation. Uhlig et al. [148] and Rocaries and Rigaud [149] report a dual-logarithmic growth for NiO in the range 307–442°C. Hauffe et al. [150] interpret their results for nickel in terms of a logarithmic law followed by a fourth-power rate law.

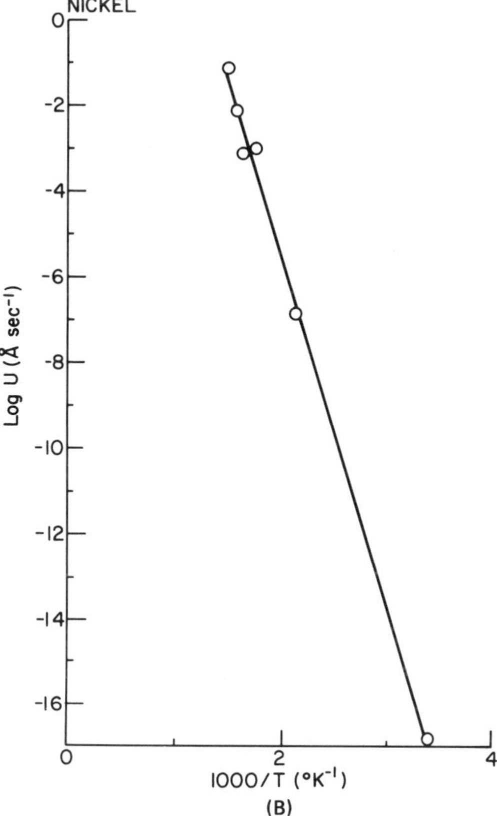

Figure 8.8. (*Continued*)

The effect of crystallographic orientation as shown in Figure 6.6 has also been explained in terms of the presence of paths of easy-ion movement in nickel oxide. This is discussed by Cathcart et al. [151].

Two different sets of data points were analyzed. The 23–340°C data (Figure 8.8A) fit both direct and inverse logarithmic kinetics with equally high (>0.970) correlation coefficients. Further analysis of the inverse log data using $Z = 2$, $2a = 4.2$ Å, and $\Omega = 16.6$ (Å)^3 (cubic NiO) gave the following results: $W = 1.6$ eV, $N = 2 \times 10^{13}$ cm^{-2}, and $|V| = 2.6$ V. The correlation coefficient of log u versus $1/T$ was 0.998 while that of x_1 versus $1/T$ was 0.957.

The data between 23 and 400°C were also analyzed. The 450°C curve

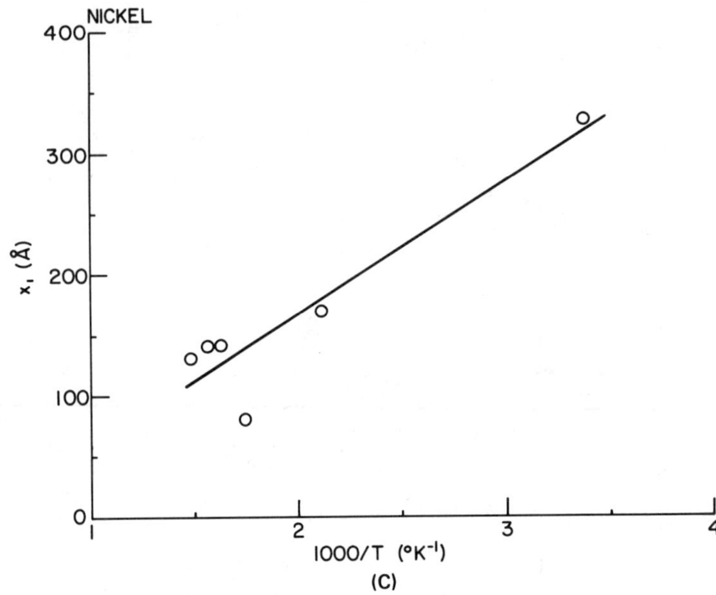

Figure 8.8. (*Continued*)

was omitted since it obviously did not fit the inverse log kinetics. Correlation coefficients were slightly higher for the inverse log (>0.970) than for the direct log (>0.910) kinetics. Analysis of inverse log results gave $W = 1.6$ eV, $N = 3 \times 10^{13}$ cm^{-2}, and $|V| = 2.3$ V. Correlation coefficients were 0.933 for x_1 (Figure 8.8C) and 0.998 for log u (Figure 8.8B). The value of W may be compared with that of 1.8 eV found by Graham and Cohen [142] when they analyzed the higher temperature results according to parabolic kinetics. Surface potential measurements [127] showed a change of -0.95 V when oxygen was adsorbed on a nickel film (25°C, 10^{-2} Torr).

8.2.3.8. ***Tantalum.*** Vermilyea's [152] data for the anodization of tantalum between 150 and 300°C (Figure 8.9A) were utilized by Ghez [153] as his prime example illustrating the Cabrera–Mott law.

Values of $W = 1.6$ eV and $N = 3 \times 10^{12}$ cm^{-2} were derived from the data (Figure 8.9B) taking $\Omega = 84$ (Å)^3 (rhombohedral Ta_2O_5). The value of W may be compared with that obtained at higher temperatures, 1.1 eV as found for the oxidation range 250–450°C at 0.1 atm where parabolic oxidation yields an adherent oxide [154]. Values of W equal to 0.71 eV

[2b] and 2.185 eV [2c] had been reported based on fitting data to empirical rate equations for anodization.

Ghez calculated a value of $|V| = 1.79$ V using $Z = 5$ and $2a = 1.5$ Å (see Figure 8.9C). However, assuming anion conductivity, Z may be taken to be 2. With $2a = 4.38$ Å, the value of $|V|$ becomes 1.5 V.

8.2.3.9. Copper. The data of Rhodin [155] were analyzed at four temperatures between −195 and 50°C as reported for the (100), (110), and (111) faces of a copper crystal. See Figure 6.3. Isotherms had been recorded by Rhodin over the course of one month and included data for two typical runs made under each set of experimental conditions. No pressure dependence was noted in the range 10^{-2}–10^{2} Torr oxygen. The relative oxidation rates were (100) > (110) > (111).

Plots of $(1/x)$ versus log (t/x^2) had correlation coefficients ranging from 0.35 to 0.99. Much of the scatter was due to combining the dual runs. Unfortunately, no sensible values of W, N, or $|V|$ could be determined from this data.

Similar data of Young et al. [156] were also plotted. See Figure 6.4. Isotherms for the oxidation of (100), (111), (110), and (311) faces of a copper single crystal had been reported for five temperatures between 70 and 178°C. The relative rates were (100) > (111) > (110) > (311). Care had been taken to prepare a clean, strain-free copper surface since contaminants greatly affected the (110) rate relative to the (111) rate.

Plots of both x versus log t and $(1/x)$ versus log (t/x^2) showed excellent correlation coefficients. On this basis, there would be little reason to choose one set of kinetics over the other.

Values of W and N calculated from the plots of log u versus $1/T$ (correlation coefficients for the crystal faces varied between 0.77 and 0.96) were both reasonable for copper oxidation, except the high value $N = 2 \times 10^{18}$ cm^{-2} for (111) copper. This is greater than a monolayer. It is noteworthy that Young et al. fitted the data from 70 to 178°C to an inverse logarithmic plot for only the (311) and (110) faces. Their qualitative fit was not confirmed by more quantitative studies. The lack of agreement was ascribed to imperfections in the oxide.

Values of voltage were too large and the correlation coefficients for x_1 versus $1/T$ too small (0.29 to 0.46). Hence $|V|$ was not considered to be correct for copper oxidation.

The data sets of Rhodin and Young et al. covered two different temperature ranges, so that, despite the scatter in Rhodin's data, it seemed logical to try combining both sets of data. Figure 8.10A contains the combined result for log u versus $1/T$ while Figure 8.10B contains x_1 versus $1/T$.

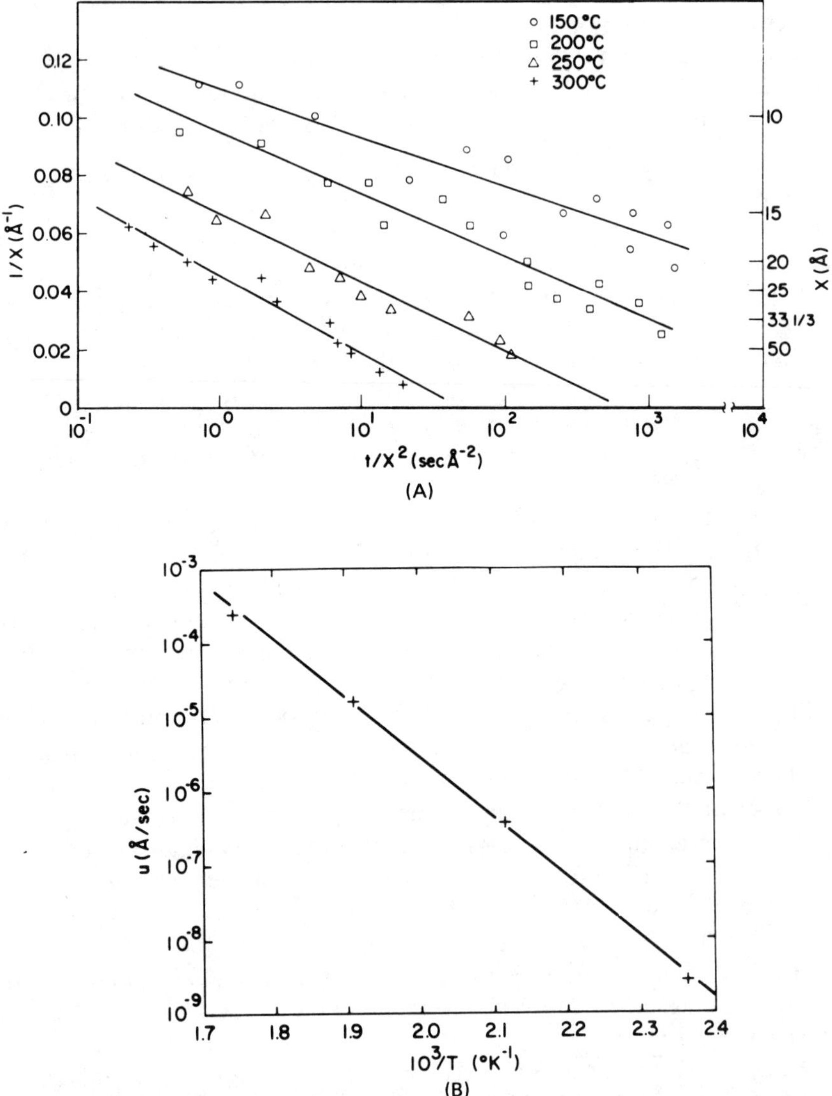

Figure 8.9. Tantalum oxidation. (A) Anodic oxide growth on tantalum sheet. (B) Plot of log u versus $1/T$. (C) Plot of x_1 versus $1/T$. After Ghez [153]. Published by permission of the copyright holder, the American Institute of Physics.

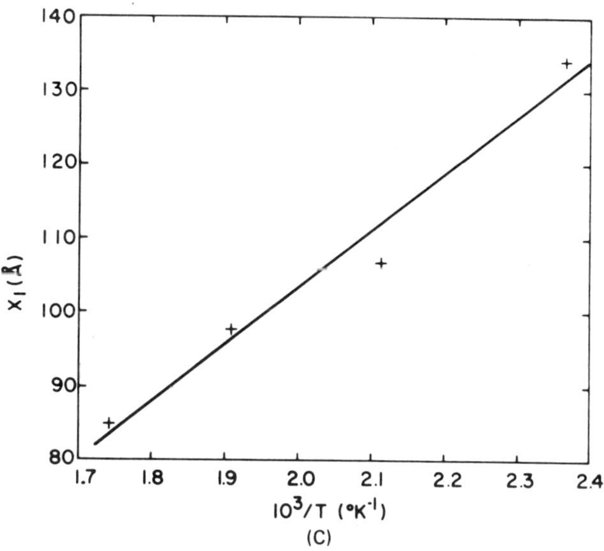

Figure 8.9. (*Continued*)

In Figure 8.10A, the data for all crystal faces group together as a function of temperature except those obtained at -195°C. The plot was used to obtain values of $W = 1.9$ eV and $N = 1 \times 10^{21}$ cm^{-2} taking $\Omega = 39.6$ (Å)^3 (cubic Cu_2O). The value of W can be compared with those obtained for parabolic oxidation in air [157]; 0.87 eV for 300–550°C and 1.63 eV for 550–900°C. Moisture has some effect on the rate of copper oxidation [46] so the comparison is only approximate.

It was apparent that the value of N found by combining the two data sets was too large since the number of sites is $\sim 10^6$ times that of a monolayer.

A value for $|V|$ could not be obtained from Figure 8.10B since the scatter in the points made it difficult, if not impossible, to define a straight line that would cut the Y-axis even when the points at -195°C were omitted. The implication is that the combined data for copper do not fit the Cabrera–Mott rate law.

Copper was investigated further by analyzing the data of Campbell and Thomas [46]. See Figure 8.11A. They measured the rate of both wet and dry oxidation between 100 and 256°C. Oxygen pressure was 150 Torr with 15 Torr added moisture in the wet runs. They found that kinetics appeared to follow a cubic rate law.

The data for dry oxidation were plotted according to both the direct

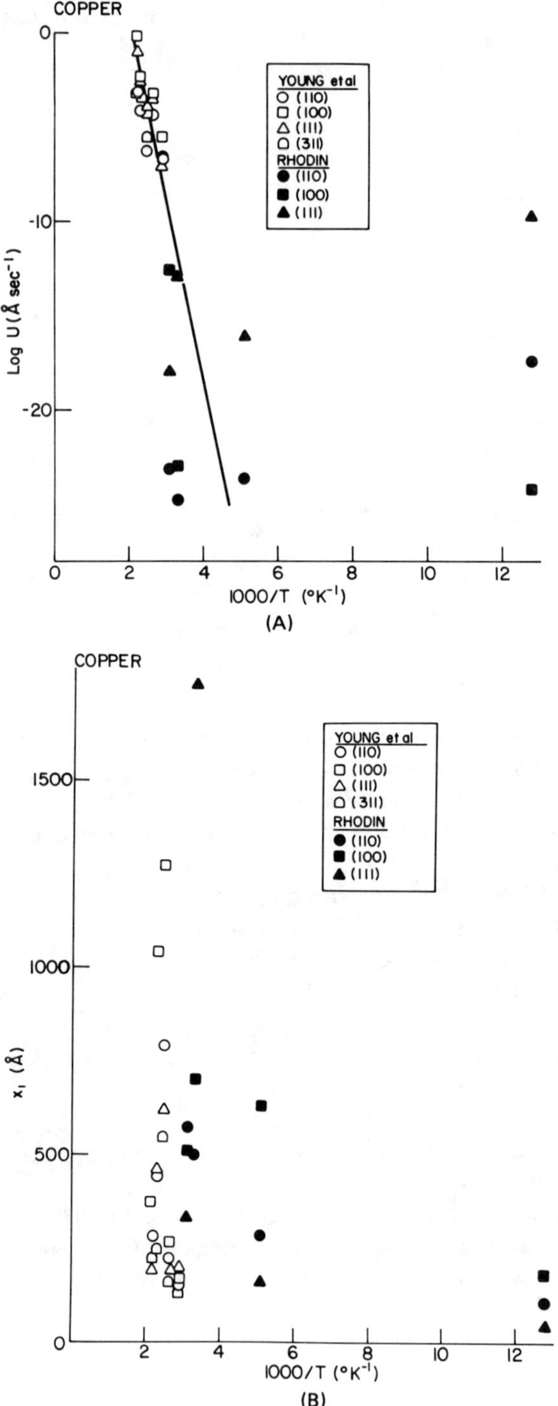
COPPER
Log U (Å sec⁻¹)
YOUNG et al
(110)
(100)
(111)
(311)
RHODIN
(110)
(100)
(111)
1000/T (°K⁻¹)
(A)
COPPER
x₁ (Å)
YOUNG et al
(110)
(100)
(111)
(311)
RHODIN
(110)
(100)
(111)
1000/T (°K⁻¹)
(B)

(Figure 8.11B) and inverse (Figure 8.11C) logarithmic rate laws. While correlation coefficients were >0.920 in both cases, they were higher for the direct log plots so no values of N or W were calculated. In addition, values of voltage calculated from the inverse log data were too high. This was also true of wet oxidation between 100 and 169°C.

The overall conclusion from the three sets of data on copper oxidation is that direct rather than inverse logarithmic kinetics better fit both the wet and dry low-temperature oxidation data.

8.2.3.10. Sodium. The dry oxidation rate of sodium was measured between −79 and 48°C by Cathcart et al. [158] as shown in Figure 8.12A. A manometric method was used at ~200-Torr pressure. Carefully purified sodium films on glass formed the substrate. Oxide film structure was not determined, but logarithmic-type kinetics were followed.

Analysis according to the direct and inverse logarithmic kinetics gave good correlation coefficients for the direct log plot (>0.957 except for −20°C which was 0.885) but poor ones for the indirect log plot (<0.572 except for 35°C which was 0.821). Attempts to calculate values of W, N, and $|V|$ gave impossible results as shown in Table 8.2. Values of $Z = 1$, $2a = 5$ Å, and $\Omega = 22.7$ (Å)^3 were used in the analysis.

It was concluded that direct log kinetics (Figure 8.12B) are a better fit than inverse log kinetics (Figure 8.12C).

8.2.3.11. Beryllium. D. E. Fowler [159] measured the rate of oxidation of a Be (0001) crystal face. An ultrahigh vacuum apparatus was used in the temperature range 27–497°C at oxygen pressures of 3×10^{-8} to 2×10^{-6} Torr. Adsorption of oxygen on the clean Be surface was followed by oxide island nucleation and growth. The islands coalesced into an oxide film which grew according to direct logarithmic kinetics. At high temperatures, BeO (0001) epitaxy on Be (0001) was found. As temperatures decreased, epitaxy became progressively poorer, being unobservable at 27°C.

Reconstruction of the BeO (0001) surface was observed at a critical thickness of 6 to 7 monolayers of oxide. This was attributed to intrinsic polarization of the wurtzite structure of BeO. Electron enhancement of oxidation by high intensity Auger beams was found and minimized.

Figure 8.10. Copper oxidation for the combined results of Rhodin [155] and Young et al. [156] on single crystal copper faces. (A) Plot of log u versus $1/T$. (B) Plot of x_1 versus $1/T$. After Fehlner [96]. Repinted by permission of the publisher, The Electrochemical Society, Inc.

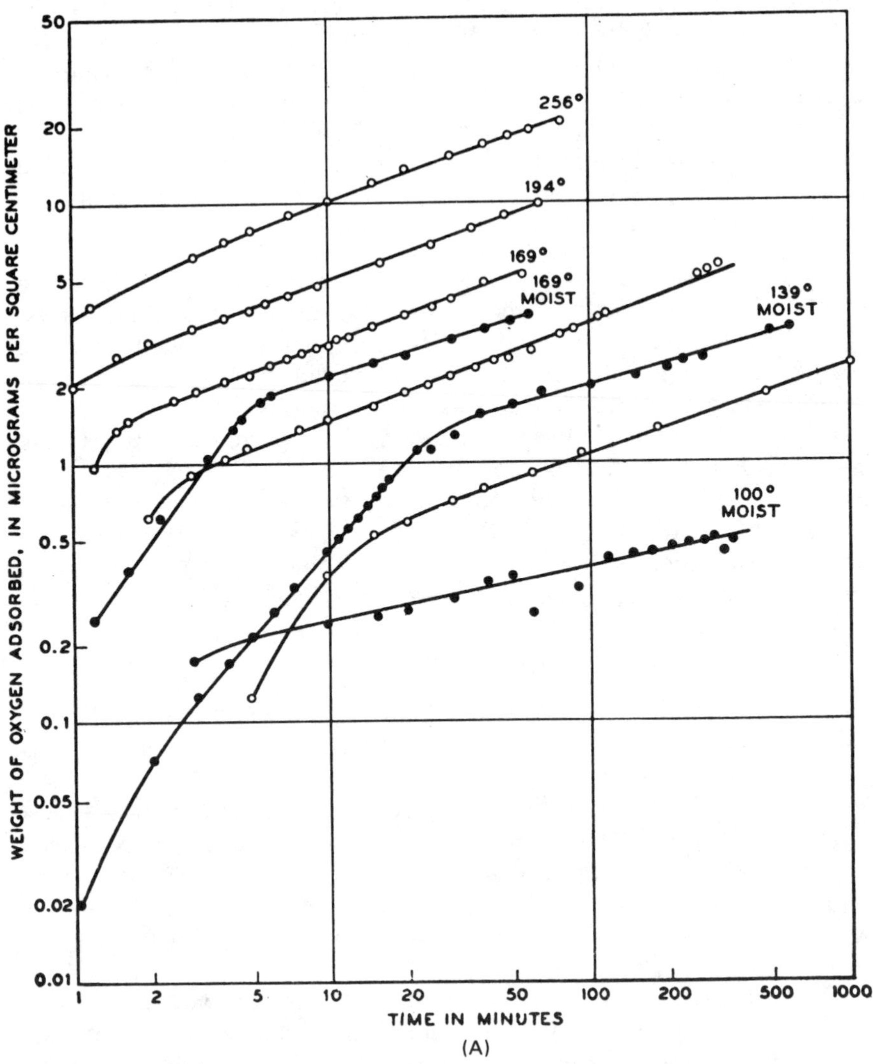

Figure 8.11. Copper oxidation. (A) Oxide growth on OFHC copper foil as a function of temperature under dry (open circles) and wet (solid circles) conditions. One μg cm^{-2} O_2 is equivalent to 149 Å Cu_2O. After Campbell and Thomas [46]. Reprinted by permission of the publisher, The Electrochemical Society, Inc. (B) Dry data plotted according to direct logarithmic kinetics. (C) Dry data plotted according to inverse logarithmic kinetics.

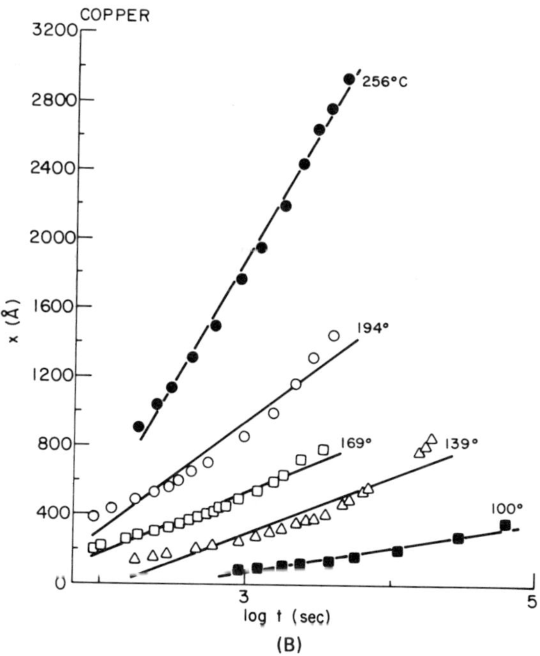

(B)

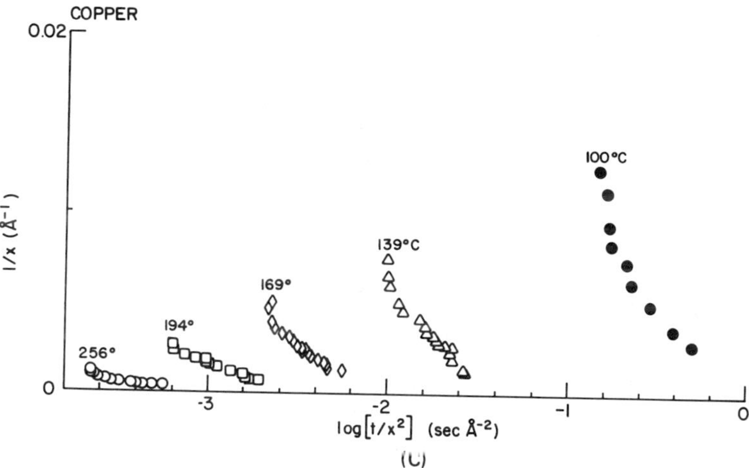

(C)

Figure 8.11. *(Continued)*

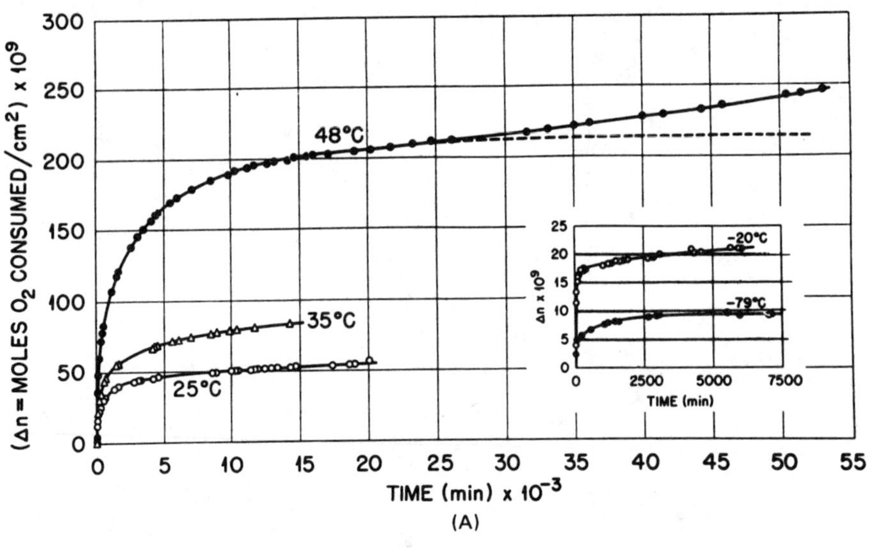

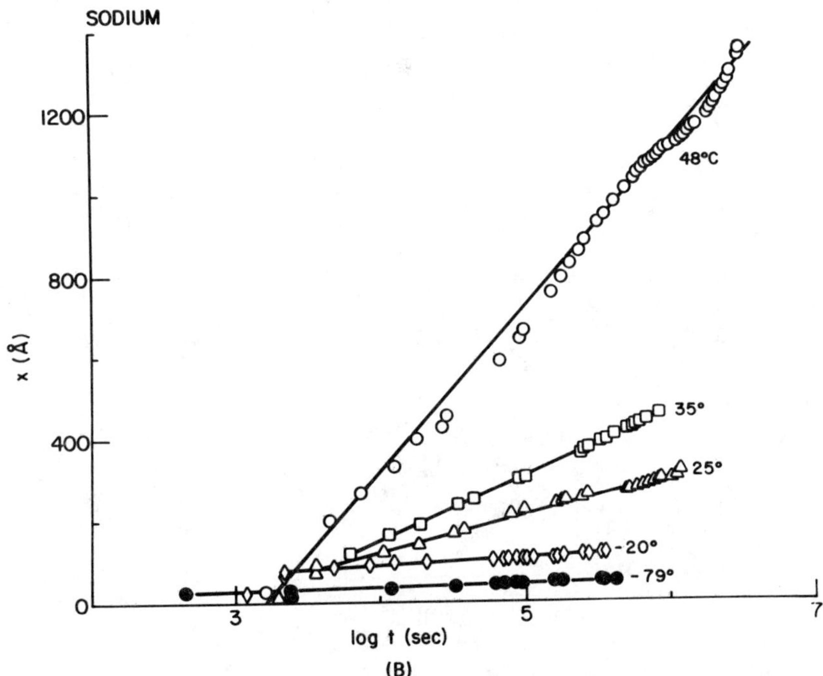

Figure 8.12. Sodium oxidation. (A) Oxide growth on sodium films as a function of temperature where 10^{-7} moles cm^{-2} O_2 is equivalent to 546 Å Na_2O. Reprinted with permission from *Acta Metallurgica*, Vol. 5, p. 247, J. V. Cathcart, L. L. Hall, and G. P. Smith, The Oxidation Characteristics of the Alkali Metals, Copyright 1957, Pergamon Press, Ltd. (B) Direct logarithmic plot of the data. (C) Inverse logarithmic plot of the data.

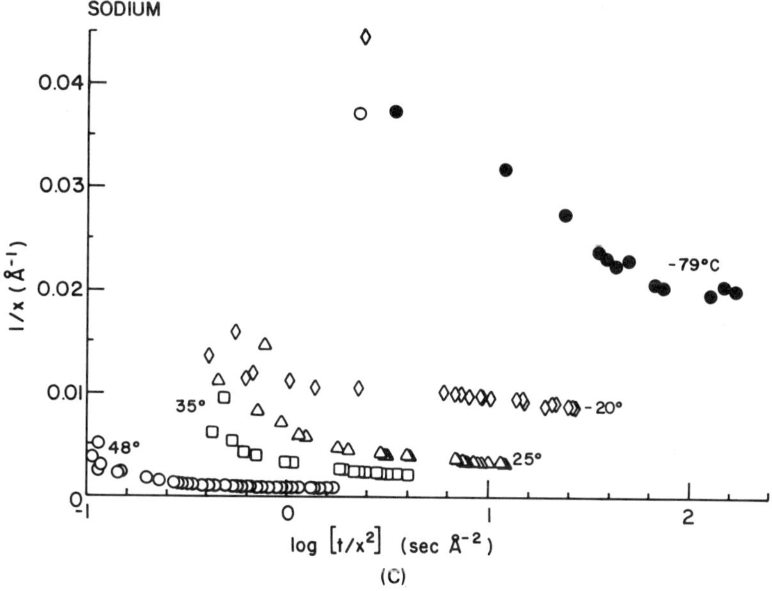

Figure 8.12. (*Continued*)

Data (Figure 8.13A) for two different maximum exposures (1.05×10^4 L and 4.2×10^3 L) were analyzed according to both the direct and inverse logarithmic rate laws. Pressure was 10^{-6} Torr. In both cases, it was obvious that the direct log kinetics fit the data while inverse log did not. The former (Figure 8.13B) gave straight lines with good correlation coefficients (>0.990 except for one 27°C curve which was 0.867) while the latter (Figure 8.13C) produced L-shaped plots with poor correlation coefficients. Fowler pointed out that both the island and continuous oxide growth stages were included in this data, but even when the island stage was removed, direct logarithmic kinetics still fit better. No attempt was made to calculate values of W, N, and $|V|$ in view of this preference for direct log kinetics.

8.2.4. Discussion

The results of the above analyses are collected together in Table 8.2. The metals are arranged according to the cation group numbers of the periodic table. Values of N, W, and $|V|$ are included.

The first observation is that, in general, the results found for the Group I and II metals are unrealistic and in some cases physically impossible.

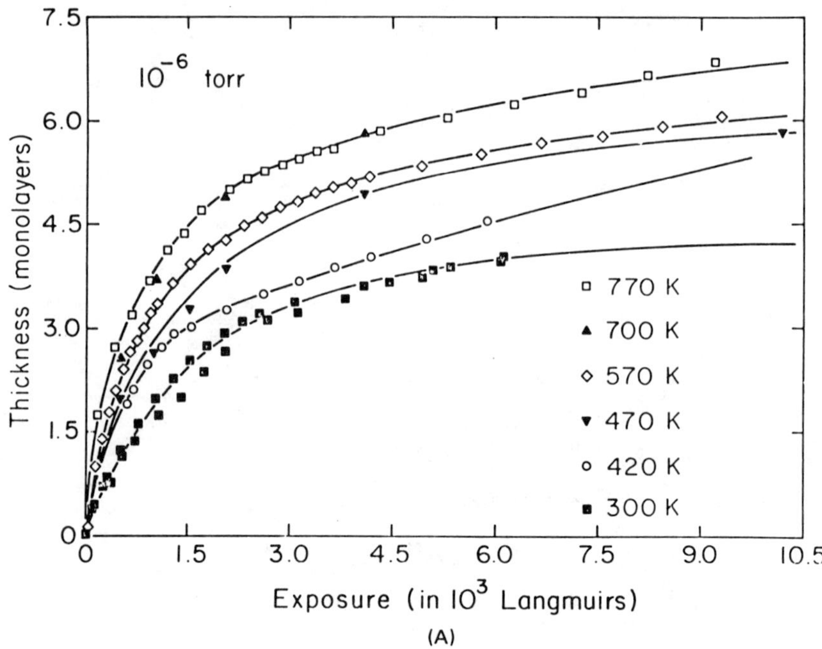

Figure 8.13. Beryllium oxidation. (A) Oxide growth on (0001) Be as a function of temperature. One monolayer is equivalent to 2.2 Å BeO. After Fowler [159]. Published by permission of the author. (B) Direct logarithmic plot of the data. (C) Inverse logarithmic plot of the data.

As a result, they should not be included in the table. For instance, one value of N is too large for Cu and too small for Na, beyond the limits of error which are estimated to be an order of magnitude. Beryllium is a similar case and no values of N and W were even calculated. On the other hand, values of N, W, and $|V|$ for the other metals in Table 8.2 are reasonable and it appears that the Cabrera–Mott model is applicable.

A comparison of the various values of N (excluding Groups I and II) shows an interesting correlation. The value, in general, decreases from Group III to IV and then increases again for Group V through VIII. Two exceptions are found: the second set of results for aluminum, and the results for iron. Oxides of these metals are considered to be intermediate oxides in the sense that they can be incorporated into, as well as modify a network. Both anion and cation movement are found in such oxides. This flexibility suggests that such oxides may shift roles depending on the conditions of oxidation.

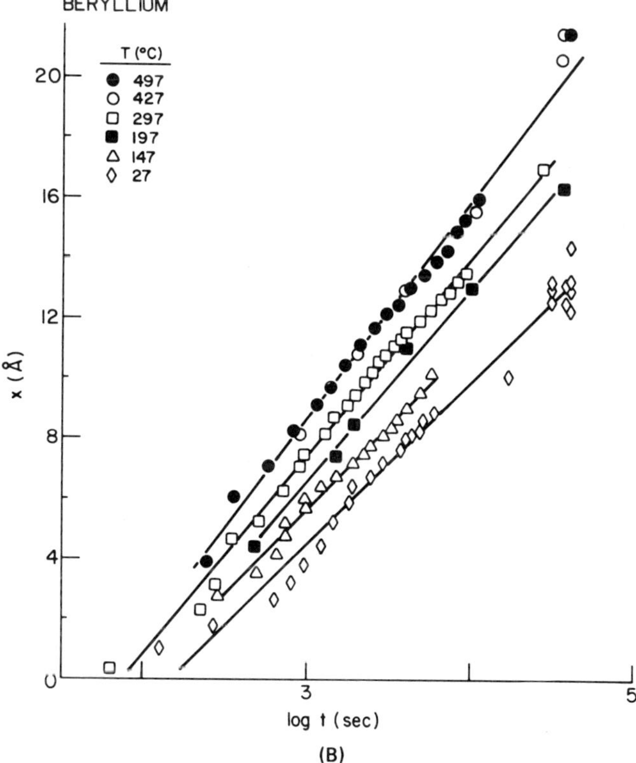

Figure 8.13. *(Continued)*

The overall trend appears to correlate with ionic transport numbers for metal oxides as listed in Table 4.1 taken mainly from work on anodic oxides [160]. Cation transport predominates for the Group I and II oxides, but quickly diminishes as the group number increases toward Group IV. Here, anion movement predominates. As the group number increases still further, cation movement again becomes significant.

Oxide structure, especially at the interfaces, is the key to understanding ion movement [43,44]. It is noteworthy that the oxides of Group I and II metals are classified as network modifiers, that is, they break up the three-dimensional network of a vitreous oxide. At room temperature and above, films of these oxides tend to be polycrystalline. The movement of small cations is favored by the presence of defects in the close-packed ionic solid.

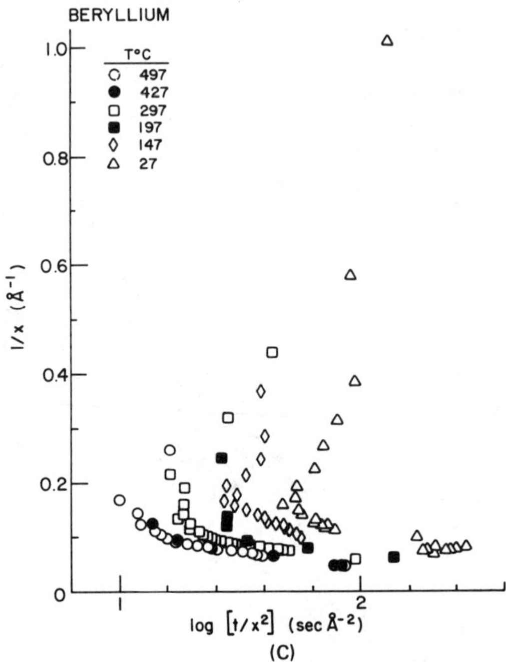

Figure 8.13. (*Continued*)

The direct logarithmic kinetics of Group I and II metals may indicate that such oxides do not fit the Cabrera–Mott model. Young and Dignam [126], however, point out "that failure of data to adhere to an inverse logarithmic law cannot be interpreted as implying the inapplicability of the Mott–Cabrera model". This conclusion is based on a comparison of calculated curves for the sinh, inverse, and direct logarithmic laws.

An alternative model for direct logarithmic kinetics has not yet been proven, although there are two possibilities as discussed below. The important points of difference between these modifiers and the network-forming or intermediate oxides which better fit Cabrera–Mott inverse logarithmic kinetics are the polycrystalline structure of the oxide and the predominance of cation movement.

The direct logarithmic law of Evans (Equation 34 of Chapter 2) would be applicable to modifiers if small crystals initially formed, giving rise to many paths of easy-ion movement. Recrystallization of the oxide during subsequent growth would decrease the number of grains. As a result, the number of paths for ion movement along grain boundaries would decrease.

An alternative explanation for copper oxidation has been based on the electronic properties of the oxide. Traps in the oxide gradually charge up, restricting further flow of electrons needed for charge neutralization at the oxide–gas interface. Uhlig assumed two space charge regions could exist in the oxide, leading to the dual-logarithmic equation (Equations 36 and 37 of Chapter 2). Williams and Hayfield [161] modified Uhlig's treatment by assuming a single, uniform space charge. Their result was a direct logarithmic expression. See Equations 38–40 in Chapter 2. These equations form the basis for Roy and Sircar's [162] treatment of copper oxidation at 75–100°C. They calculated that the number of trapped electrons was $\sim 10^{18}$ cm^{-3} and the flux of Fermi electrons moving toward the metal–oxide barrier equaled 4×10^{16} cm^{-2} sec^{-1}.

The oxides of the elements in Group IV are classified as network formers. They are capable of bonding into a three-dimensional network which can change the mechanism of ionic self-diffusion. The strong covalent bonding in the network promotes formation of holes and channels through which large anions or even oxygen molecules can pass. Cations, because of their high charge density, are less mobile in such a network structure.

Given the network oxide structure and anion conductivity, it is not surprising that N derived from the Cabrera–Mott expression is small. This is so because N can no longer refer to the potentially mobile cation population at the metal–oxide interface. Instead, it is proposed that it refers to the number of potentially mobile anions at the oxide–gas interface which in turn may relate to the number of preferred paths available for ion movement through the oxide. It follows that where N is relatively small and W is >1 eV, the rate of oxide growth is also small. Thus, the addition of alloying elements to base metals to promote network oxide formation at the surface leads to protective oxide films and corrosion resistance, for example, Cr in stainless steel [44].

Network-forming oxides tend to be vitreous at room temperature. The temperatures at which the oxide network remains stable to recrystallization are determined by the nature of the cation, the presence of impurities, and applied stresses. The best network-forming oxide, SiO_2, when grown on silicon [163], is stable at high temperatures, in the 1200°C range.

Oxides of metals in Groups III, V, VI, and VIII are intermediates. As a result, their values of N are expected to be larger than those of network formers. The disagreement in the values of N for aluminum may relate to the fact that both cations and anions can move in this intermediate oxide. The relative ratio of the two is known to vary with anodization conditions and can also vary with conditions of thermal oxidation, for example, pressure, and the nature of the metal substrate, whether it is foil [115,116] or film [63].

Barr [164] studied the passivation in air of elemental metals using ESCA. His results could be divided into two general classes. In the first, passivation terminated with the outermost layer of metal oxide, that is, that exposed to air, in its highest oxidation state, but with the bulk of the metal oxide in a lower oxidation state. Such oxides were classified as modifiers according to the nomenclature of glass chemists.

The second class of oxides terminated in the same oxidation state throughout the oxide. These oxides were classed as network formers. The results were complicated by the presence of water, but the correlation with the present work is impressive. For example, the network formers were Si, Al, Zr, Sn, Mo, W, Y, La, and Rh while the modifiers were Ce, Pt, Pd, Fe, Co, Ni, and Cu.

The application of the Cabrera–Mott theory to both cation- and anion-conducting oxides is a question discussed by Fehlner and Mott [43]. It is noteworthy that in the case of cation movement, oxidation kinetics are affected more by crystalline orientation of the metal substrate than by oxygen pressure. The converse is true for anion movement.

8.2.5. Conclusions

The present results are limited in scope, especially as regards modifying oxides. Nevertheless, it appears that oxidation according to the Cabrera–Mott theory occurs when a uniform, stable, three-dimensional oxide film is formed. This occurs at different temperatures dependent on the ability of the metal to form vitreous oxide.

Metals which form modifying oxides having a polycrystalline structure do not appear to fit the Cabrera–Mott inverse logarithmic kinetics. Instead, they follow direct logarithmic kinetics.

Values of W obtained from Cabrera–Mott analysis of kinetic data are similar to values obtained from high-temperature oxidation. This implies that a similar rate-limiting step, perhaps bond breaking, may control oxidation over the whole span of temperature.

Values of N correlate with the structure and mode of ion transport in oxides. It is smallest for network formers where anion movement predominates. A larger value is found for intermediates where both anion and cation movement occur.

Consistent with the above, values of the absolute value of voltage V are reasonable except for modifiers. Comparison with values determined from surface potential measurements shows fairly good agreement.

REFERENCES

1. N. Cabrera and N. F. Mott, *Rep. Prog. Phys.* **12** (1948–49), 163.
2. L. Young, *Anodic Oxide Films*, Academic Press, New York, 1961: (a) p. 220, (b) p. 93, (c) p. 99.
3. D. D. Davis and D. L. Johnson, *J. Electrochem. Soc.* **125** (1978), 1889.
4. C. J. Dell'Oca, D. L. Pulfrey, and L. Young, in *Physics of Thin Films*, Vol. 6, M. H. Francombe and R. W. Hoffman, Eds., Academic Press, New York, 1971, p. 1.
5. J. H. Greiner, *J. Appl. Phys.* **42** (1971), 5151.
6. A. T. Fromhold, Jr. and M. Baker, *ibid.* **51** (1980), 6377.
7. P. J. Jorgensen, *J. Chem. Phys.* **37** (1962), 874.
8. J. L. Miles, *Proc. Roy. Soc.* (*London*) **A321** (1971), 503.
9. R. C. Jaklevic, *J. Electrochem. Soc.* **126** (1979), 1548.
10. J. Perriere, J. Siejka, and M. Croset, *ibid.* **128** (1981), 530.
11. C. Y. Yang and W. E. O'Grady, *J. Vac. Sci. Technol.* **20** (1982), 925, 1056.
12. D. O. Raleigh, *J. Electrochem. Soc.* **113** (1966), 782; **115** (1968), 111.
13. A. T. Fromhold, Jr., *Theory of Metal Oxidation, Vol. 1—Fundamentals*, North-Holland, New York, 1976.
14. S. S. Gill and I. H. Wilson, *Thin Solid Films* **55** (1978), 435.
15. L. A. Kasprzak and A. K. Gaind, *IBM J. Res. Develop.* **24** (1980), 348.
16. O. Kubaschewski and B. E. Hopkins, *Oxidation of Metals and Alloys*, 2nd ed., Butterworks, London, 1962: (a) p. 168ff, (b) p. 259.
17. *Microweighing in Vacuum and Controlled Environments*, A. W. Czanderna and S. P. Wolsky, Eds., Elsevier, Amsterdam, 1980.
18. G. Sauerbrey, *Z. Physik* **155** (1959), 206.
19. V. Mecea and R. V. Bucur, *J. Vac. Sci. Technol.* **17** (1980), 182.
20. E. A. Gulbransen, *ibid.*, 109.
21. *Ellipsometry in the Measurement of Surfaces and Thin Films*, E. Passaglia et al., Eds., NBS Misc. Publ. 256, U.S. Department of Commerce, Washington, DC, 1964.
22. *Ellipsometry* (Proc. 4th Intern'l Conf. Ellipsometry, Berkeley, 1979), *Surface Sci.* **96** (1980), Nos. 1–3.
23. J. L. Ord, *Surface Sci.* **16** (1969), 155.
24. T. Smith, P. Smith, and F. Mansfeld, *J. Electrochem. Soc.* **126** (1979), 799.
25. G. A. Swallow and G. C. Allen, *Oxid. Metals* **17** (1982), 141.
26. G. C. Allen and G. A. Swallow, *ibid.*, 157.
27. H. H. Uhlig, *Corrosion and Corrosion Control*, 2nd ed., John Wiley & Sons, New York, 1971.
28. J. Kruger and R. P. Frankenthal, in *Techniques of Metals Research*, Vol. IV, Part 2, R. A. Rapp, Ed., John Wiley & Sons, New York, 1970, p. 571.
29. C. R. Helms, *J. Vac. Sci. Technol.* **20** (1982), 948.
30. G. Bayreuther, *ibid.* **Al** (1983), 19.
31. J. M. Khan and D. M. Makowiecki, *Surface Sci.* **77** (1978), L155.
32. *Tunneling Spectroscopy* Paul K. Hansma, Ed., Plenum Press, New York, 1982,
33. W. J. Nelson, D. G. Walmsley, and J. M. Bell, *Thin Solid Films* **79** (1981), 229.
34. R. S. Alwitt, C. K. Dyer, and B. Noble, *J. Electrochem. Soc.* **129** (1982), 711.
35. R. S. Timsit, W. G. Waddington, C. J. Humphreys, and J. L. Hutchison, *Appl. Phys. Lett.* **46** (1985), 830.
36. D. K. Ottesen and A. S. Nagelberg, *Thin Solid Films* **73** (1980), 347.
37. R. L. Farrow, P. L. Mattern, and A. S. Nagelberg, *Appl. Phys. Lett.* **36** (1980), 212.

38. R. L. Farrow, R. E. Benner, A. S. Nagelberg, and P. L. Mattern, *Thin Solid Films* **73** (1980), 353.
39. S. G. Schultz, M. Janik-Czachor, and R. P. Van Duyne *Surface Sci.* **104** (1981), 419.
40. A. J. Melmed and J. J. Carroll, *J. Vac. Sci. Technol.* **10** (1973), 164.
41. E. E. Huber, Jr., and C. T. Kirk, Jr., *Surface Sci.* **5** (1966), 447.
42. M. W. Roberts and B. R. Wells, *Surface Sci.* **8** (1967), 453; **15** (1969), 325.
43. F. P. Fehlner and N. F. Mott, *Oxid. Metals* **2** (1970), 59.
44. A. G. Revesz and J. Kruger, in *Passivity of Metals*, R. P. Frankenthal and J. Kruger, Eds., The Electrochemical Society Inc., Princeton, NJ, 1978, p. 137.
45. R. Gomer and G. Tryson, *J. Chem. Phys.* **66** (1977), 4413.
46. W. E. Campbell and U. B. Thomas, *Trans. Electrochem. Soc.* **91** (1947), 623.
47. A. G. Revesz and F. P. Fehlner, *Oxid. Metals* **15** (1981), 297.
48. G. Okamoto and T. Shibata, *Nature* **206** (1965), 1350.
49. *Atmospheric Corrosion*, W. H. Ailor, Ed., John Wiley & Sons, New York, 1982.
50. *The Corrosion Handbook*, H. H. Uhlig, Ed., John Wiley & Sons, New York, 1948.
51. M. Pourbaix, in Ref. 49, p. 107.
52. S. W. Dean, Jr., in Ref. 49, p. 195.
53. M. R. Pinnel, H. G. Tompkins, and D. E. Heath, *J. Vac. Sci. Technol.* **16** (1979), 161.
54. D. W. Rice, R. J. Cappell, P. B. P. Phipps, and P. Peterson, in Ref. 49, p. 651.
55. D. W. Rice, R. J. Cappell, W. Kinsolving, and J. J. Laskowski, *J. Electrochem Soc.* **127** (1980), 891.
56. D. W. Rice, P. Peterson, E. B. Rigby, P. B. P. Phipps, R. J. Cappell, and R. Tremoureux, *ibid.* **128** (1981), 275.
57. R. P. Frankenthal, P. C. Milner, and D. J. Siconolfi, *ibid.* **132** (1985), 1019.
58. D. W. Rice, P. B. P. Phipps, and R. Tremoureux, *ibid.* **126** (1979), 1459.
59. D. W. Rice, P. B. P. Phipps, and R. Tremoureux, *ibid.* **127** (1980), 563.
60. M. R. Pinnel, H. G. Tompkins, and D. E. Heath, *ibid.* **126** (1979), 1274.
61. H. G. Tompkins and J. A. Augis, *J. Vac. Sci. Technol.* **18** (1981), 408.
62. B. E. Hayden, W. Wyrobisch, W. Oppermann, S. Hachicha, P. Hofmann, and A. M. Bradshaw, *Surface Sci.* **109** (1981), 207.
63. J. Grimblot and J. M. Eldridge, *J. Electrochem. Soc.* **128** (1981), 729.
64. J. Grimblot and J. M. Eldridge, *ibid.* **129** (1982), 2366, 2369.
65. G. L. Hunt and I. M. Ritchie, *Oxid. Metals* **2** (1970), 361.
66. J. M. Eldridge, Y. J. van der Meulen, and D. W. Dong, *Thin Solid Films* **12** (1972), 447.
67. K. Hauffe, in *Surface Phenomena of Metals*, S.C.I. Monograph No. 28, Society of Chemical Industry, London, 1968, p. 327.
68. W. H. J. Vernon, E. I. Akeroyd, and E. G. Stroud, *J. Inst. Metals* **65** (1939), 301.
69. G. E. Hammer and R. M. Shemenski, *J. Vac. Sci. Technol.* **A1** (1983), 1026.
70. P. J. Jorgensen, *J. Electrochem. Soc.* **110** (1963), 461.
71. J. R. Anderson and I. M. Ritchie, *Proc. Roy. Soc.* (*London*) **A299** (1967), 371.
72. J. W. Lindsay, K. Terada, and M. A. Thompson, *J. Electrochem. Soc.* **119** (1972), 726.
73. L. W. Owen and J. R. Alderton, *Proc. Intern. Congr. Metal. Corrosion, 2nd, New York, 1963*, National Assoc. Corr. Eng., Houston, 1966, p. 865.
74. M. G. Hapase, V. B. Tare, and A. B. Biswas, *Acta Met.* **15** (1967), 131.
75. R. M. Tahboub, M. El Guindy, and H. D. Merchant, *Oxid. Metals* **13** (1979), 545.
76. R. W. Joyner, M. W. Roberts, and S. P. Singh-Boparai, *Surface Sci.* **104** (1981), L199.
77. L. H. Lee and V. Y. Doo, *J. Electrochem. Soc.* **118** (1971), 443.
78. T. Smith, *Surface Sci.* **38** (1973), 292.

79. A. Milch and P. Tasaico, *J. Electrochem. Soc.* **127** (1980), 884.
80. R. G. Musket, *Surface Sci.* **74** (1978), 423.
81. R. R. Addiss, Jr., *Acta. Met.* **11** (1963), 129.
82. B. Kasemo and L. Walldén, *Surface Sci.* **75** (1978), L379.
83. E. H. Nicollian and J. R. Brews, *MOS (Metal-Oxide-Semiconductor) Physics and Technology*, John Wiley & Sons, New York, 1982.
84. F. P. Fehlner, *J. Electrochem. Soc.* **119** (1972), 1723.
85. B. Carrière, A. Chouiyakh, and B. Lang, *Surface Sci.* **126** (1983), 495.
86. F. Lukeš, *ibid.* **30** (1972), 91.
87. R. J. Archer, *J. Electrochem. Soc.* **104** (1957), 619.
88. G. Mende, J. Finster, D. Flamm, and D. Schulze, *Surface Sci.* **128** (1983), 169.
89. S. I. Raider, R. Flitsch, and M. J. Palmer, *J. Electrochem. Soc.* **122** (1975), 413.
90. F. P. Fehlner, *ibid.*, 1745.
91. J. P. Ponpon and B. Bourdon, *Solid-State Electronics* **25** (1982), 875.
92. S. R. Kelemen, Y. Goldstein, and B. Abeles, *Surface Sci.* **116** (1982), 488.
93. A. C. Adams, T. E. Smith, and C. C. Chang, *J. Electrochem. Soc.* **127** (1980), 1787.
94. M. Hirayama, H. Miyoshi, N. Tsubouchi, and H. Abe, *J. Electron. Mater.* **11** (1982), 919.
95. A. Cros, J. Derrien, and F. Salvan, *Surface Sci.* **110** (1981), 471.
96. F. P. Fehlner, *J. Electrochem. Soc.* **131** (1984), 1645.
97. J. R. Ligenza, *J. Phys. Chem.* **64** (1960), 1017.
98. N. V. Mileshkina and R. Z. Bakhtizin, *Surface Sci.* **29** (1972), 644.
99. Y. Kamigaki and Y. Itoh, *J. Appl. Phys.* **48** (1977), 2891.
100. M. Horiuchi, Y. Kamigaki, and T. Hagiwara, *J. Electrochem. Soc.* **125** (1978), 766.
101. M. A. Hopper, R. A. Clarke, and L. Young, *ibid.* **122** (1975), 1216.
102. A. M. Goodman and J. M. Breece, *ibid.* **117** (1970), 982.
103. B. E. Deal and A. S. Grove, *J. Appl. Phys.* **36** (1965), 3770.
104. A. G. Revesz and H. A. Schaeffer, *J. Electrochem. Soc.* **129** (1982), 357.
105. L. N. Lie, R. R. Razouk, and B. E. Deal, *ibid.*, 2828.
106. L. Pauling, *The Nature of the Chemical Bond*, 3rd ed., Cornell University Press, Ithaca, New York, 1960, p. 85.
107. E. A. Irene and Y. J. van der Meulen, *J. Electrochem. Soc.* **123** (1976), 1380.
108. G. Hollinger and F. J. Himpsel, *J. Vac. Sci. Technol.* **A1** (1983), 640.
109. J. M. Eldridge and D. W. Dong, *Surface Sci.* **40** (1973), 512.
110. J. M. Eldridge, *ibid.*, 531.
111. N. J. Chou, J. M. Eldridge, R. Hammer, and D. Dong, *J. Electronic Mater.* **2** (1973), 115.
112. T. F. Archbold and R. E. Grace, *Trans. Met. Soc. AIME* **212** (1958), 658.
113. J. R. Anderson and V. B. Tare, *J. Phys. Chem.* **68** (1964), 1482.
114. J. M. Saleh, B. R. Wells, and M. W. Roberts, *Trans. Faraday Soc.* **60** (1964), 1865.
115. M. J. Dignam, W. R. Fawcett, and H. Böhni, *J. Electrochem. Soc.* **113** (1966), 656.
116. M. J. Dignam, *ibid.* **109** (1962), 184, 192.
117. M. J. Pryor, *Oxid. Metals* **3** (1971), 271.
118. R. W. Bartlett, *J. Electrochem. Soc.* **111** (1964), 903.
119. R. L. Wells and T. Fort, Jr., *Surface Sci.* **33** (1972), 172.
120. C. T. Kirk, Jr. and E. E. Huber, Jr., *ibid.* **9** (1968), 217.
121. G. L. Hunt and I. M. Ritchie, *J. Chem. Soc. Faraday Trans. I* **68** (1972), 1413.
122. W. Eberhardt and C. Kunz, *Surface Sci.* **75** (1978), 709.
123. D. J. Young and M. Cohen, *J. Electrochem. Soc.* **124**, (1977), 769, 775.
124. K. Shanker and P. H. Holloway, *Thin Solid Films* **105** (1983), 293.

125. G. A. Hope and I. M. Ritchie, *ibid.* **34** (1976), 111.
126. D. J. Young and M. J. Dignam, *J. Phys. Chem. Solids* **34** (1973), 1235.
127. C. M. Quinn and M. W. Roberts, *Trans. Faraday Soc.* **60** (1964), 899.
128. W. A. Crossland and H. T. Roettgers, *Phys. Failure Electron.* **5** (1966), 158.
129. E. A. Gulbransen and K. F. Andrew, *J. Electrochem. Soc.* **104** (1957), 334.
130. M. Cohen, *ibid.* **121**, (1974), 191C.
131. K. Kuroda, B. D. Cahan, Gh. Nazri, E. Yeager, and T. E. Mitchell, *ibid.* **129** (1982) 2163.
132. M. Langell and G. A. Somorjai, *J. Vac. Sci. Technol.* **21** (1982), 858.
133. W. G. Dorfeld, J. B. Hudson, and R. Zuhr, *Surface Sci.* **57** (1976), 460.
134. M. J. Graham, S. I. Ali, and M. Cohen, *J. Electrochem. Soc.* **117** (1970), 513.
135. M. J. Graham and M. Cohen, *ibid.* **116** (1969), 1430.
136. P. B. Needham, Jr., H. W. Leavenworth, Jr., and T. J. Driscoll, *ibid.* **120** (1973), 778.
137. P. B. Sewell, C. D. Stockbridge, and M. Cohen, *ibid.* **108** (1961), 933.
138. N. H. Turner, R. J. Colton, and J. S. Murday, *J. Vac. Sci. Technol.* **18** (1981), 593.
139. J. Kruger and H. T. Yolken, *Corrosion* **20** (1964), 29t.
140. K. Bhavani and V. K. Vaidyan, *Oxid. Metals* **15** (1981), 137.
141. P. B. Abel, A. H. Heuer, and R. W. Hoffman, *J. Vac. Sci. Technol.* **A1** (1983), 260.
142. M. J. Graham and M. Cohen, *J. Electrochem. Soc.* **119** (1972), 879.
143. M. J. Graham, G. I. Sproule, D. Caplan, and M. Cohen, *ibid.* **119** (1972), 883.
144. S. Mróz, C. Koziol, and J. Kolaczkiewicz, *Vacuum* **26** (1976), 61.
145. D. F. Mitchell, P. B. Sewell, and M. Cohen, *Surface Sci.* **69** (1977), 310.
146. D. F. Mitchell, P. B. Sewell, and M. Cohen, *ibid.* **61** (1976), 355.
147. W. J. Moore and J. K. Lee, *Trans. Faraday Soc.* **48** (1952), 916.
148. H. Uhlig, J. Pickett, and J. MacNairn, *Acta Met.* **7** (1959), 111.
149. J. C. Rocaries and M. Rigaud, *Scripta Met.* **5** (1971), 59.
150. K. Hauffe, L. Pethe, R. Schmidt, and S. Roy Morrison, *J. Electrochem. Soc.* **115** (1968), 456.
151. J. V. Cathcart, G. F. Petersen, and C. J. Sparks, Jr., *ibid.* **116** (1969), 664.
152. D. A. Vermilyea, *Acta Met.* **6** (1958), 166.
153. R. Ghez, *J. Chem. Phys.* **58** (1973), 1838.
154. E. A. Gulbransen and K. F. Andrew, *J. Metals, N.Y.* **188** (1950), 586.
155. T. N. Rhodin, Jr., *J. Amer. Chem. Soc.* **72** (1950), 5102; **73** (1951), 3143.
156. F. W. Young, Jr., J. V. Cathcart, and A. T. Gwathmey, *Acta Met.* **4** (1956), 145.
157. G. Valensi, *Rev. mét.* **45** (1948), 205.
158. J. V. Cathcart, L. L. Hall, and G. P. Smith, *Acta Met.* **5** (1957), 245.
159. D. E. Fowler, The Initial Stages of Oxidation of the Beryllium Single Crystal Surface (0001), Ph.D. Thesis, Cornell University, Ithaca, May, 1983.
160. J. P. S. Pringle, *Electrochim. Acta* **25** (1980), 1403, 1423.
161. E. C. Williams and P. C. S. Hayfield, in *Vacancies and Other Point Defects in Metals and Alloys*, Monograph No. 23, The Institute of Metals, London, 1958, p. 131.
162. S. K. Roy and S. C. Sircar, *Oxid. Metals* **15** (1981), 9.
163. W. H. Dumbaugh and P. C. Schultz, *Encyclopedia of Chemical Technology*, 2nd ed., Vol. 18, John Wiley & Sons, New York, 1969, p. 73.
164. T. L. Barr, *J. Phys. Chem.* **82** (1978), 1801.

9

OXIDATION OF SILICON

9.1. INTRODUCTION

The conversion of crystalline silicon to vitreous silicon dioxide during thermal oxidation is the outstanding example illustrating the role of vitreous oxides in oxidation. The role of impurities [1] such as hydroxyls [2] and substrate perfection [3] are critical to the perfection of the resulting dielectric which forms part of all integrated circuits. As such, vitreous silica has received a great deal of attention. The study of metal-oxide-semiconductor (MOS) devices [4] provides us with a wealth of data on charged states, both in the oxide and at the oxide–silicon interface. It is possible, as a result, to gain some insight into the structure of the oxide and the effects of processing variables on this structure.

Vitreous silica, as the noncrystalline form of silicon dioxide is called, is widely used in industries other than electronics. As a result, several confusing terms have evolved. Defining these may be of help to the reader.

Dumbaugh and Schultz [5] point out that "vitreous silica" is the most unambiguous term. "Silica glass" is similar, but sometimes describes oxide glasses in which SiO_2 is the major component of a mixture. "Fused silica" describes the process as much as the product. Here, the vitreous silica is formed by methods other than melting of crystalline quartz, for example, by flame hydrolysis of $SiCl_4$. "Fused quartz" is, as the name implies, produced by direct melting of quartz crystals. In the present work we shall adopt vitreous silica as the appropriate term for the noncrystalline oxide.

Noncrystalline SiO_2 films formed on oxidized silicon resemble vitreous silica in their properties and structure [6,7]. The oxide may be thought of as an open network containing interstices a few angstroms in diameter and ~20 Å apart [7,8]. Gas molecules readily dissolve and diffuse [9] through this structure. (See Chapter 4.) However, under certain conditions the network may be changed. For instance, Revesz and Schaeffer

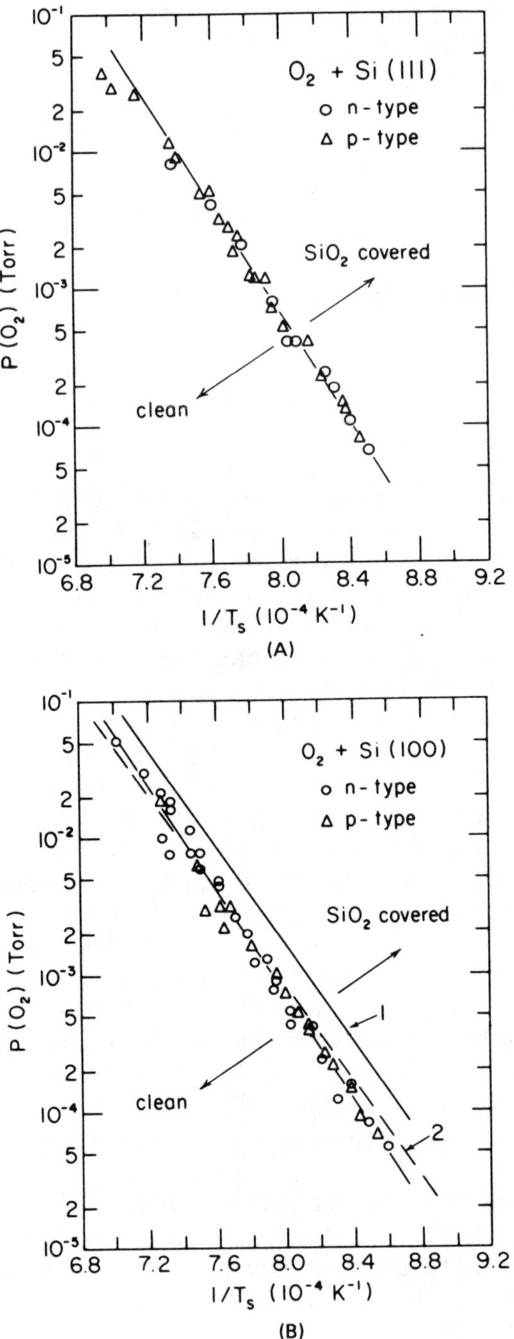

O2 + Si (111)
n-type
p-type
SiO2 covered
clean
P (O2) (Torr)
1/Ts (10^-4 K^-1)
(A)
O2 + Si (100)
n-type
p-type
SiO2 covered
clean
1
2
P (O2) (Torr)
1/Ts (10^-4 K^-1)
(B)

[8] discuss the possible role of structural channels formed during the early stages of oxidation under dry conditions. Barr [10] interprets his ESCA results in terms of a strained region in the silica near the Si/SiO_2 interface following a model put forward by Grunthaner and Maserjian [11]. Grunthaner et al. [12] interpret their XPS results in terms of an ~20 Å-thick strained interfacial layer. This is structurally but not stoichiometrically different from the bulk oxide.

Critical conditions of temperature, pressure, and surface preparation are required for the growth of vitreous SiO_2 on silicon [13]. The volatile suboxide SiO can form and evaporate at reduced pressures as found by Derrien and Commandré [14]. Figure 9.1 shows the conditions of temperature and pressure under which SiO_2 forms. Oxygen pressure in this case equals total pressure. If the oxygen partial pressure is reduced by adding an inert gas such as nitrogen while maintaining total pressure at, say, 1 atm, then the lines in Figure 9.1 will move toward lower pressures for fixed temperature [15].

Oxidation can also be carried out in wet oxygen or pyrogenic steam. Water reacts with the SiO_2 network to break it up, making reactant movement easier. Thus, the rate of oxide growth is higher than in dry oxygen. However, at sufficiently high temperatures and pressures, silica begins to dissolve in water and the overall rate of oxide growth decreases [16].

The growth of vitreous silica from silicon also depends upon conditions of extreme cleanliness [4]. The usual temperatures of oxidation, 900–1200°C, lie in the range of silica devitrification. This is a process in which the vitreous structure rearranges to form cristobalite. This crystalline phase is usually nucleated at a surface which has been sensitized by impurities such as Na^+ or Al^{3+} [17]. An oxygen deficiency in the glass tends to inhibit devitrification [18].

The structure of vitreous silica is discussed in Chapter 3. Silica exists as a covalent compound having a three-dimensional random network such that almost all the bonds are complete. Silicon atoms are tetrahedrally surrounded by oxygen atoms. Bond lengths are fixed at values characteristic of the crystalline compound. However, the O–Si–O angles can vary from 120° to 180°, giving the structure flexibility. Transport phe-

←

Figure 9.1. Critical conditions of oxygen pressure and reciprocal substrate temperature for the growth of SiO_2 by the reaction of O_2 with Si (111) in (A) and Si (100) in (B) for both *n*- and *p*-type substrates. In (B), the equilibrium vapor pressure for SiO produced in the reaction $Si_{(s)} + SiO_{2(s)} \rightarrow 2SiO_{(g)}$ is shown by curve 1 while the prediction of a thermodynamic model for the critical growth conditions is given by curve 2. After Smith and Ghidini [13]. Reprinted by permission of the publisher, The Electrochemical Society, Inc.

nomena in silicates are such that permeability of gases takes place faster than movement of modifying ions, which in turn, diffuse at a faster rate than network ions.

9.2. KINETICS OF SILICON OXIDATION

9.2.1. Deal–Grove Model

The classic paper on the thermal oxidation of single crystal silicon is that of Deal and Grove [19]. The phenomenological model they presented fits their data for wet and dry oxidation, giving values of physical parameters which agree with those derived from other experiments. Further investigations of silicon oxidation under dry [20,21,22] and wet [23] conditions fit the same model up to pressures of 20-atm steam [24] or dry oxygen [25]. A change in activation energy, previously reported by Irene and Dong [26] as a curvature in the Arrhenius plots, was found at ~900°C in both steam and dry O_2 [24,25].

The assumptions of the Deal–Grove model are as follows, referring to Figure 9.2.

1. A species of oxidant moves from the oxide–gas interface through the oxide to react at the oxide–silicon interface. Rosencher et al. [27] have experimentally supported this interfacial reaction for dry oxidation where

$$Si + O_2 \rightarrow SiO_2 \tag{1}$$

A relatively small exchange of oxygen atoms between O_2 and SiO_2 occurs at the oxide–gas interface during this process [28]. Cristy and Condon [29] attribute this to vacancy movement on the oxygen "sublattice" of the SiO_2 phase. Breed and Doremus [30] discuss an interfacial reaction for wet oxidation

$$Si + 2H_2O \rightarrow SiO_2 + 2H_2 \tag{2}$$

Reaction with the oxide itself occurs in the case of steam oxidation

$$Si\text{–}O\text{–}Si + H_2O \rightleftarrows 2Si\text{–}OH \tag{3}$$

$$Si\text{–}O\text{–}Si + H_2 \rightleftarrows Si\text{–}OH + SiH \tag{4}$$

The process is a dynamic one.

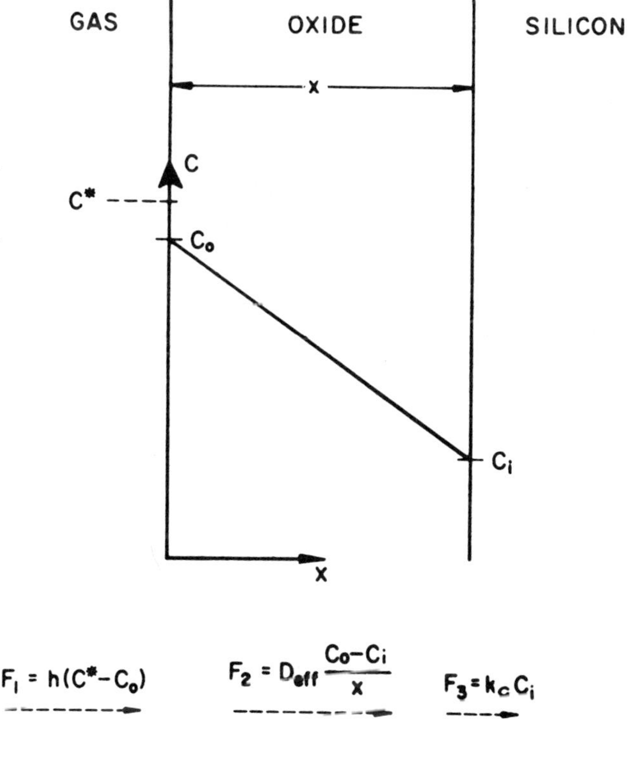

Figure 9.2. Model for the oxidation of silicon, where: C^* is the equilibrium concentration of the oxidant at the oxide surface; $C^* = K_H P$, where P is pressure and K_H is a constant. C_o is the oxidant concentration at the outer oxide surface at any given time. C_i is the oxidant concentration at the inner oxide surface. h is the gas-phase transport coefficient for oxygen. k_c is the linear rate constant for interface reaction. D_{eff} is the effective diffusion coefficient within the oxide. x is the oxide thickness. F is the flux of oxidant. After Deal and Grove [19]. Published by permission of the copyright holder, the American Institute of Physics.

2. Henry's law holds in the gas phase.

$$C^* = K_H P \tag{5}$$

3. The flux of oxidant from the gas phase into the oxide surface is

$$F_1 = h(C^* - C_o) \tag{6}$$

It is assumed that no dissociation of O_2 takes place at the oxide surface.

4. The flux of oxidant through the oxide occurs by diffusion under the influence of a concentration gradient

$$F_2 = \frac{-D_{\mathrm{eff}}(C_o - C_i)}{x} \tag{7}$$

5. The reaction of the oxidant with the silicon surface is first order as given in Equation 1 above

$$F_3 = k_c C_i \tag{8}$$

6. The three fluxes are equal in a steady-state process after growth of an initial oxide thickness x_i on the silicon.

$$F_1 = F_2 = F_3 = F \tag{9}$$

This assumption allows C_i and C_o to be eliminated from the resulting expression such that

$$F = \frac{k_c C^*}{1 + k_c/h + k_c x/D_{eff}} \tag{10}$$

The rate of oxide growth then becomes

$$\frac{dx}{dt} = \frac{F}{N_1} \tag{11}$$

where N_1 is the number of oxidant molecules incorporated into a unit volume of the oxide layer.

Given that $x = x_i$ at $t = 0$, then the solution to Equation 11 is the linear-parabolic equation originally derived by Evans [31]

$$x^2 + Ax = B(t + \tau) \tag{12}$$

where

$$A \cong 2D_{\mathrm{eff}} \left(\frac{1}{k_c} + \frac{1}{h}\right) \tag{13}$$

$$B \cong \frac{2D_{\mathrm{eff}} C^*}{N_1} \tag{14}$$

$$\tau \cong \frac{x_i^2 + Ax_i}{B} \tag{15}$$

The introduction of initial time τ converts the initial oxide thickness x_i into a shift along the time coordinate.

A characteristic time for the process is defined by

$$\frac{A^2}{4B} = \frac{D_{\text{eff}} N_1}{2C^*(1/k_c + 1/h)^2} \tag{16}$$

and a characteristic oxide thickness by

$$\frac{A}{2} = \frac{D_{\text{eff}}}{1/k_c + 1/h} \tag{17}$$

For times or oxide thicknesses less than the above, Equation 12 reduces to a linear rate law.

$$x \cong \frac{B}{A}(t + \tau) \tag{18}$$

where $t \ll A^2/4B$.

For times or oxide thicknesses larger than the characteristic ones, a parabolic law is found

$$x^2 \cong Bt \tag{19}$$

where $t \gg A^2/4B$ and $t \gg \tau$.

In this way a linear rate constant

$$\frac{B}{A} = \frac{k_c h}{k_c + h}\left(\frac{C^*}{N_1}\right) \tag{20}$$

and a parabolic rate constant as given in Equation 14 are derived. From these, values of C^*, k_c, and h can be determined based on experimental data.

9.2.2. Experimental Rates

Thermal oxidation is the usual method employed to grow SiO_2 films on silicon. Experimental results measured to $\pm 2\%$ (Figures 9.3 and 9.4) during the 1960s were found to fit Equation 12 as shown in Figure 9.5. Here, Deal and Grove [19] combined the work of four sets of authors. Both wet and dry results are included. The conclusion at that time was that Equation 12 could be used to calculate values of oxide thickness from 700 to 1300°C

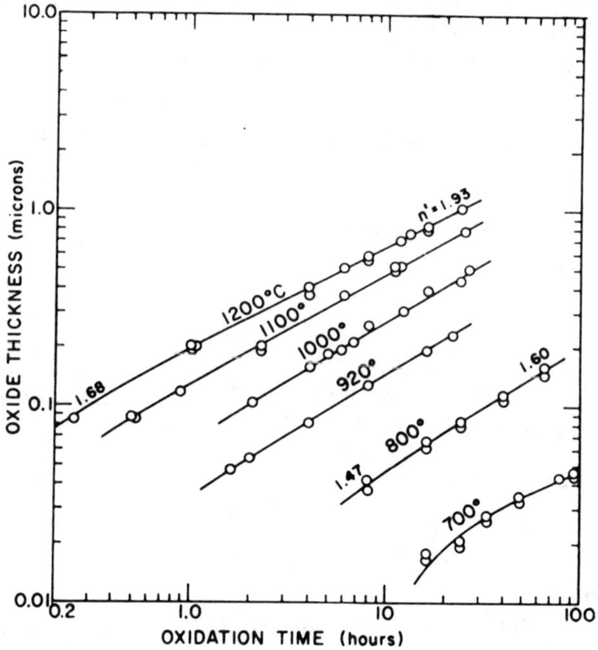

Figure 9.3. Oxidation of (111) Si in dry oxygen at 760 Torr as a function of temperature. Here n' is refractive index. After Deal and Grove [19]. Published by permission of the copyright holder, the American Institute of Physics.

under wet or dry conditions. More recent data, discussed below, indicate that further refinement of the model is possible.

The temperature and pressure dependence of the rate constants is the key to connecting silica structure with kinetic data. Papers on the high-pressure oxidation of silicon in dry oxygen at pressures up to 20 atm [25] and at 130 atm [32] as well as in pyrogenic steam [24] at pressures up to 20 atm update the experimental results of Deal and Grove. As previously found, initial oxide thickness x_i is a function of the ambient conditions. The initial value of 200 Å for dry oxidation was remeasured to be 40 Å. The value for wet oxidation remained zero except at low temperatures. Revesz [33] comments on the significance of these findings as they affect the possible existence of channels in SiO_2.

Overall rate of oxidation varies with crystal face, (111) giving faster oxide growth than (100) both in dry O_2 (Figure 9.6) and steam (Figure 9.7). This difference shows up in a variation with crystal face of the linear rate constant B/A rather than a variation in B, the parabolic rate constant.

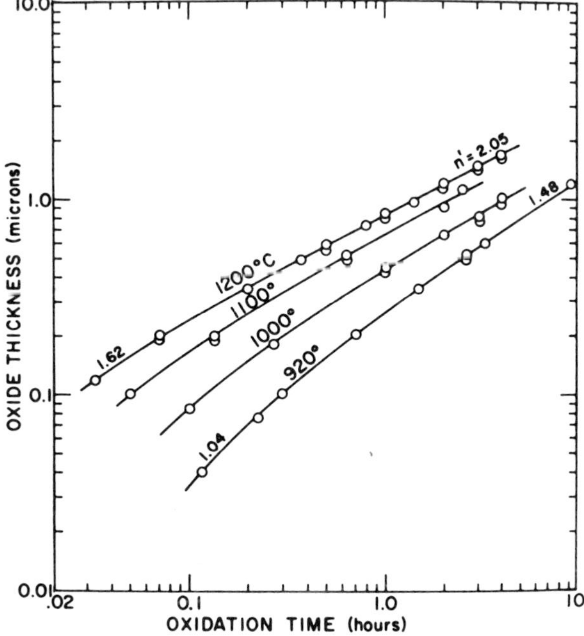

Figure 9.4. Oxidation of (111) Si in wet (95°C H_2O) oxygen at 760 Torr as a function of temperature. Here n' is refractive index. After Deal and Grove [19]. Published by permission of the copyright holder, the American Institute of Physics.

Such results agree with the Deal–Grove model in which B/A relates to a reaction at the Si/SiO_2 interface while B relates to diffusion through a homogeneous oxide. There is a reversal of relative rates at low oxygen pressures such that the (100) face supports a higher rate of oxidation than the (111) face [34].

9.2.3. Activation Energies

Activation energies for both B and B/A were determined under wet and dry conditions. Although the data were sparse, temperature dependence of the linear rate constant B/A was interpreted in terms of activation energies of 1.87–1.89 eV for dry and 1.95–2.05 eV for wet oxidation. The closeness of these values was taken to indicate that a similar interfacial reaction characterized by rate constant k_c occurred in both cases. It was suggested that breaking of a Si–Si bond is the key event since the Si–Si bond energy is 1.83 eV [35]. See Figures 9.8B and 9.9B.

The activation energy for B appeared to undergo a change at 900°C.

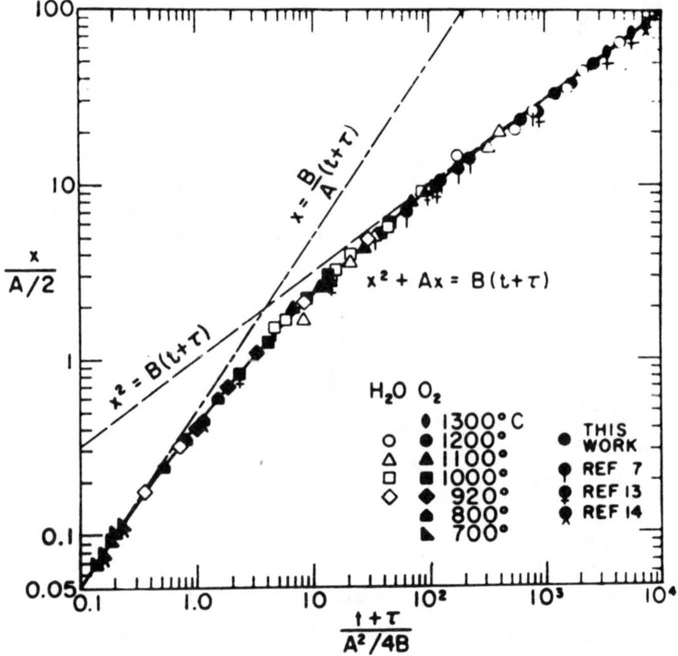

Figure 9.5. General relationship for the thermal oxidation of silicon. The solid line represents the general relationship, the dotted lines its two limiting forms. The values of x_i are taken to be zero and 200 Å for wet and dry oxygen, respectively. References can be found in the work of Deal and Grove. After Deal and Grove [19]. Published by permission of the copyright holder, the American Institute of Physics.

Under dry conditions, it was 1.77 eV between 800 and 900°C and 1.42 eV from 900–1000°C (1 to 20 atm). See Figure 9.8A. These values compare with 1.24 eV found previously by Deal and Grove [19] at 1 atm. From Equation 14, B is expected to follow the temperature dependence of the diffusivity of oxygen in fused silica. Norton [36] found a value of 1.17 eV for oxygen diffusivity, somewhat lower than the values quoted above for SiO_2 film formation.

The activation energy for parabolic oxidation in steam undergoes a similar change at 900°C, from 1.17 to 0.78 eV (1–20 atm.). See Figure 9.9A. The change is more abrupt than in dry oxidation and may relate to a change in oxide viscosity as suggested by EerNisse [37]. Alternatively, it may relate to a change in oxide density and stress, as discussed by Irene, Dong, and Zeto [38,39]. Irene [40] has published a modified linear-parabolic model. He postulates an oxygen flux in micropores combined

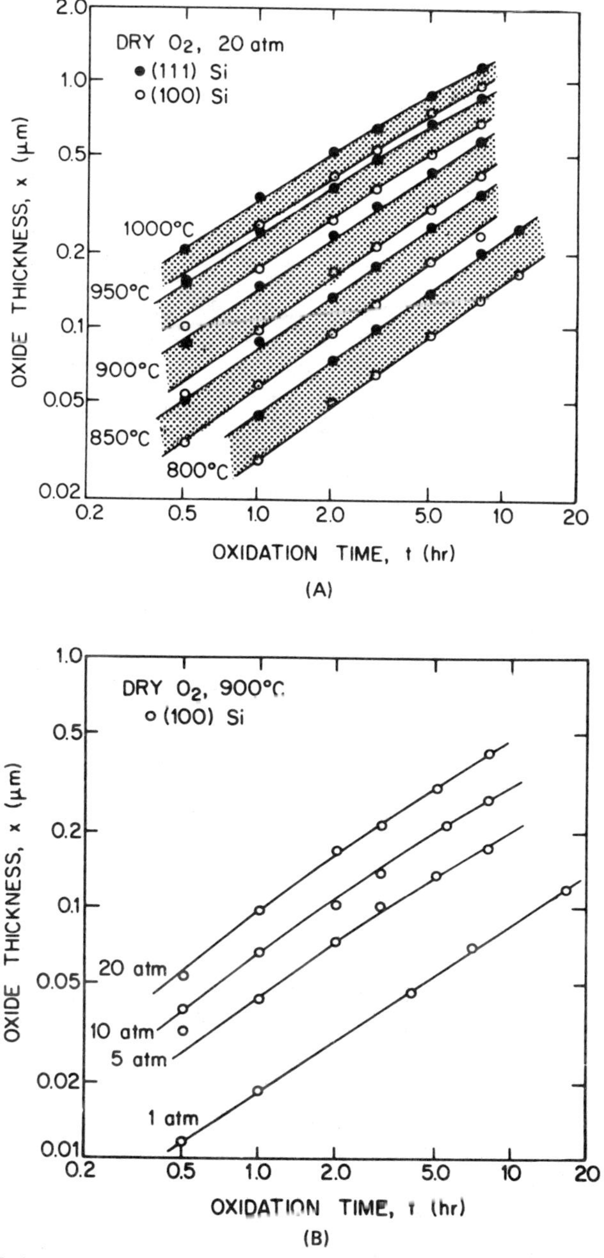

Figure 9.6. Oxide thickness versus oxidation time in dry oxygen. (A) (100) and (111) silicon oxidized at 20 atm and 800–1000°C. (B) (100) silicon oxidized at 900°C and 1, 5, 10, and 20 atm. After Lie et al. [25]. Reprinted by permission of the publisher, The Electrochemical Society, Inc.

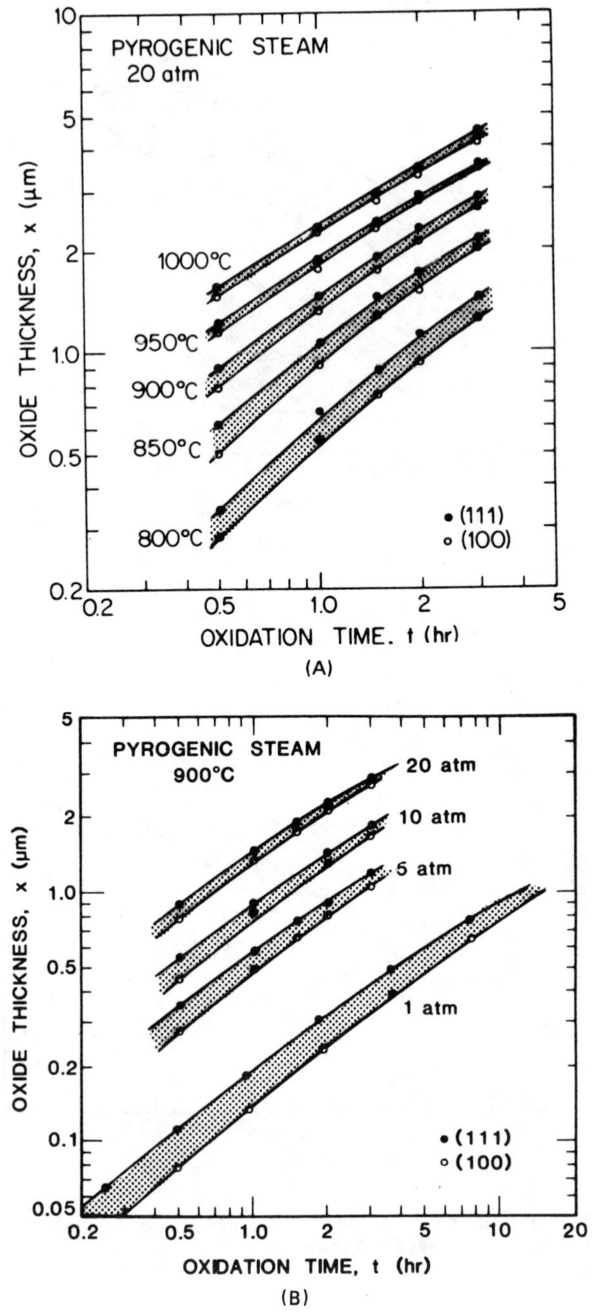
PYROGENIC STEAM
20 atm
OXIDE THICKNESS, x (μm)
1000°C
950°C
900°C
850°C
800°C
• (111)
○ (100)
OXIDATION TIME, t (hr)
(A)
PYROGENIC STEAM
900°C
20 atm
10 atm
5 atm
1 atm
OXIDE THICKNESS, x (μm)
• (111)
○ (100)
OXIDATION TIME, t (hr)
(B)

with Fickian diffusion and includes the viscoelastic properties of SiO_2. The model is used to explain high- and low-temperature oxidation, the formation of higher density SiO_2 at low temperatures, and the initial fast oxidation.

Moulson and Roberts [41] report a value of 0.79 eV for the diffusivity of water in fused silica, while Rigo et al. [42] using isotopic exchange find a value of 0.69–0.78 eV for the diffusion of OH in silica films. These values are in reasonable agreement with the value for B found by Razouk et al. [24] at temperatures above 900°C.

The fairly good agreement between values of activation energy for films and bulk silica indicates that molecular species, O_2 and/or H_2O, diffuse through the SiO_2 film. This conclusion is reinforced by the pressure dependence of the parabolic rate constants. In both dry oxygen and steam, a linear dependence is found for the range 1–20 atm. The diffusion of molecular water is further supported by the work of Breed and Doremus [30] who modeled hydrogen profiles in wet-oxidized silicon. Their work was based on the experimental results of Beckmann and Harrick [43] and Burkhardt [44]. It has also been reported that hydrogen diffuses as a molecular species in SiO_2 [45].

9.2.4. Pressure Dependence

The pressure dependence of the linear rate constant B/A was found to be proportional to P^n where $0.7 < n < 0.8$ for dry oxidation and $n = 1$ for steam. The first-order pressure dependence of both B and B/A during steam oxidation derives from Henry's law, $C^* = K_H P^n$. When n is equal to 1, k_c in B/A (Equation 20) appears to be independent of pressure, since C^* depends directly on pressure. On the other hand, it can be postulated that k_c in B/A for dry oxidation varies as P^{n-1}, in this case $P^{-0.3}$. The striking difference between dry oxygen and steam oxidation coupled with the differing values of x_i implies a difference in reaction kinetics at the Si/SiO_2 interface.

Calculations of C^*, the equilibrium concentration of reactant, O_2 or H_2O, in SiO_2 were carried out by Deal and Grove [19] taking $N_1(O_2) = 2.25 \times 10^{22}\ cm^{-3}$ and $N_1(H_2O) = 4.50 \times 10^{22}\ cm^{-3}$. Values of C^* derived

Figure 9.7. Oxide thickness versus oxidation time in pyrogenic steam. (A) (100) and (111) silicon oxidized at 20 atm as a function of temperature. (B) (100) and (111) silicon at 900°C as a function of pressure. In both cases, oxidation time includes pressurization time in steam. After Razouk et al. [24]. Reprinted by permission of the publisher, The Electrochemical Society, Inc.

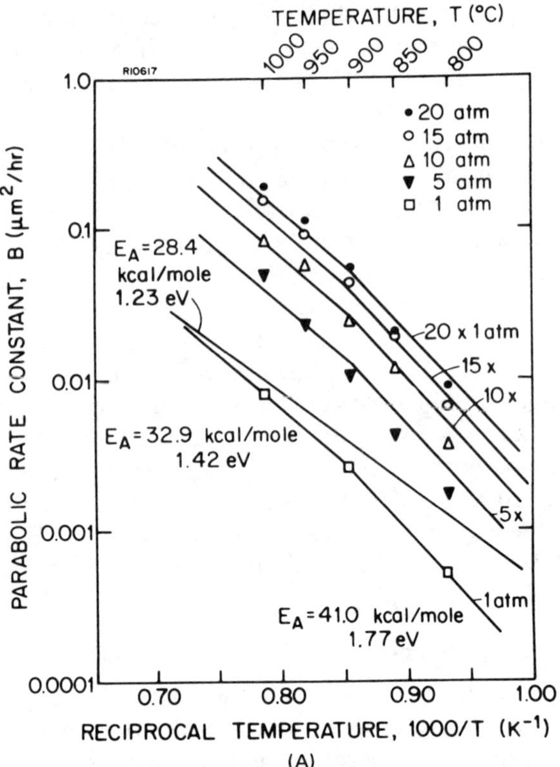

Figure 9.8. Rate constants for (111) silicon oxidation in dry oxygen. See Figure 9.6. (A) Calculated parabolic rate constants versus $1/T$ at 800–1000°C as a function of pressure. (B) Calculated linear rate constants for the same conditions as in (A). In both cases, x_i was taken to be 40 Å. After Lie et al. [25]. Reprinted by permission of the publisher, The Electrochemical Society, Inc.

by them are compared in Table 9.1 with values determined by others. The agreement is good.

Values of h and k_c were also determined. It was shown that h is large, $\sim 10^8\ \mu\text{m hr}^{-1}$ so that

$$k_c = \begin{cases} 3.6 \times 10^4\ \mu\text{m hr}^{-1} \text{ in dry } O_2 \\ 1.8 \times 10^3\ \mu\text{m hr}^{-1} \text{ in wet } O_2 \end{cases} \tag{21}$$

at 1000°C and 760 Torr.

The oxidation process was not sensitive to a fiftyfold change in carrier

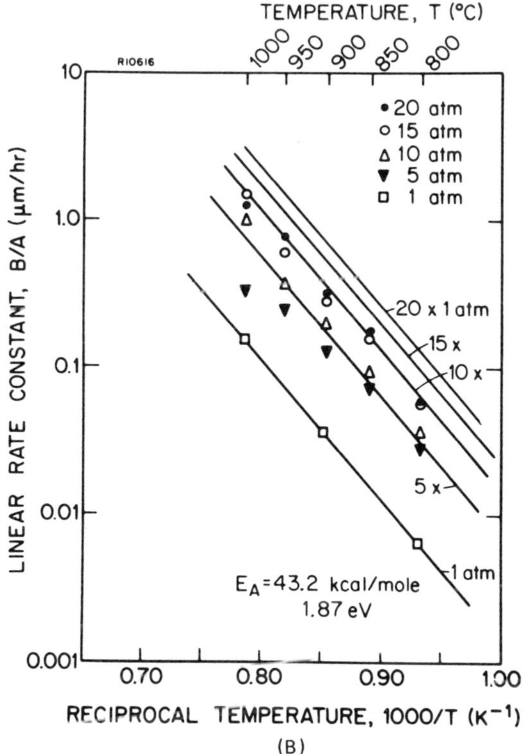

Figure 9.8. (*Continued*)

TABLE 9.1. SOLUBILITY OF OXIDIZING SPECIES IN SiO_2 AT 1000°C

Species	Solubility (cm^{-3})	
	Deal and Grove [19]	Other methods
O_2	5.2×10^{16}	5.5×10^{16} Permeation [36]
H_2O	3.0×10^{19}	3.4×10^{19} Infrared [41]

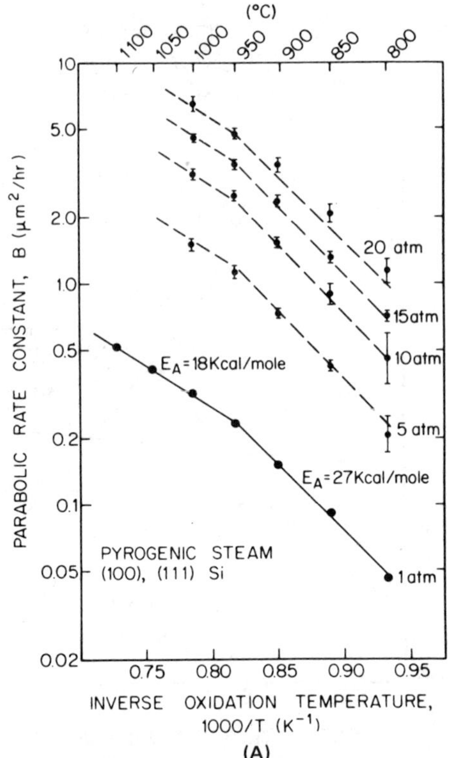

Figure 9.9. Rate constants for (100) and (111) silicon in pyrogenic steam. See Figure 9.7. (A) Calculated parabolic rate constants versus $1/T$ at 800–1100°C as a function of pressure. (B) Calculated linear rate constants versus $1/T$ for the same conditions as in (A). Calculations took into account the effect of pressurization time. After Razouk et al. [24]. Reprinted by permission of the publisher, The Electrochemical Society, Inc.

gas flow rate, helping confirm the large value of h. It is interesting that, although k_c (H_2O) is less than $k_c(O_2)$, the rate of oxidation in the presence of water is greater. This is because $C^*(H_2O)$ is larger than $C^*(O_2)$ as shown in Table 9.1.

9.2.5. Summary of Kinetics

The emphasis in the above discussion has been twofold: diffusion in the oxide is rate controlling for large oxide thicknesses while reaction at the oxide–silicon interface controls silicon oxidation for thin oxides. The two factors, however, are not mutually exclusive. In addition, other factors such as micropores and oxide viscosity should be considered in the basic

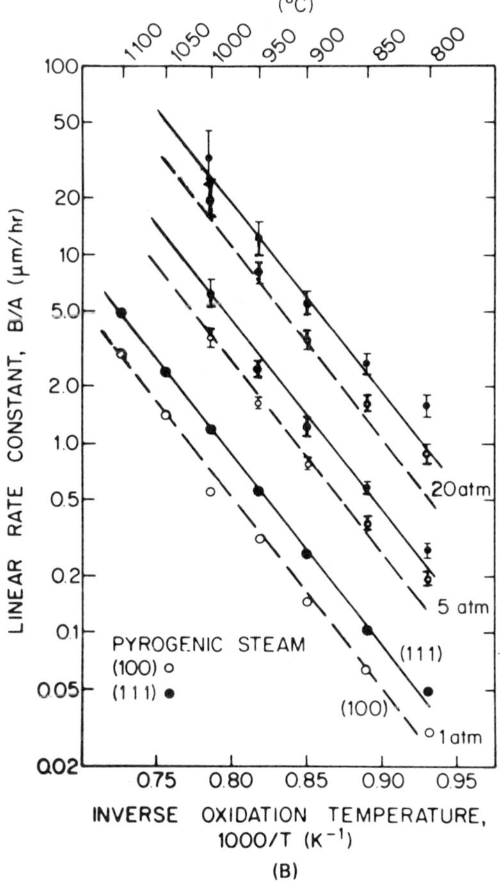

Figure 9.9. *(Continued)*

model. It is important to note that impurities such as water, sodium, or chlorine critically affect the rate of silicon oxidation as discussed by Nicollian and Brews [4]. The presence of metals such as Cu, Ag, Au, or Pd also increases the rate of silicon oxidation at low temperatures [46].

Tiller [47] has presented a generalized description of silicon oxidation in which he partitions the driving force for oxidation. His focus is on vacancy flow. A steady-state analysis leads to linear and parabolic rate constants. A network of extra half-planes is postulated to form at the Si/SiO_2 interface to reduce strain energy. Movement of the interface requires a flow of vacancies or interstitials in the silicon or free volume in the SiO_2. A two-layer model involves transformation of silicon to α-cristobalite plus silicon interstitials. Oxidation of the interstitials distorts the lattice, transforming the structure to vitreous silica. The inner layer is thin, on the order of 6 Å, and is assumed to block movement of the oxidizing species. A Deal–Grove formalism for the oxidation kinetics results from this treatment. Doremus [48] also presents a Deal–Grove formalism in which the role of strain is emphasized.

9.3. EXTENSION OF DEAL–GROVE MODEL

The work of Deal and Grove [19] left certain questions unanswered. For instance, oxide growth under the influence of applied electric fields or at very low oxygen pressures is not readily explained by their theory. Further, Irene [39] presents evidence based on lag-time diffusion measurements for different dominant modes of oxidation at higher and lower oxidation temperatures. It is particularly interesting that SiO_2 grown at lower temperatures (650–950°C) has a higher density and refractive index than that grown at 1000°C and 1 atm. Both dry oxygen [38,49] and pyrogenic steam [50] show the effect. It is independent of pressure [51]. Models based on viscous flow [52] and combined effects [40] have been formulated to modify the linear-parabolic model.

Deal and Grove did tentatively attribute values of x_i (~200 Å for dry oxidation and zero for wet) to space charge effects. However, the dominance of uncharged species as the mobile entity diffusing through the oxide at high temperatures is now accepted for the conditions of their experiments [7,8]. As a result, the role of electric fields, the growth of very thin SiO_2 films, and the structure of the oxide films require further consideration.

9.3.1. Initial Oxide Thickness

The role of x_i may be explained by considering that transport by both molecular oxygen and anions is relevant to silicon oxidation as pointed out by Revesz and Evans [20] and by Mott [53,54]. The dominance of one or the other is determined by the conditions of oxidation. Costello and Tressler [55], using isotope labeling, showed that a shift occurred from molecular transport at 1000°C to lattice oxygen diffusivity at 1300°C. The nature of the anions, whether O_2^-, O^-, O^{2-}, O_2^{2-}, or OH^-, is as yet undefined. Mott [53] reviews the evidence and even suggests that a silicon vacancy is possible. He also points out that a thermodynamic equilibrium between the neutral and charged species as postulated by Collins and Nakayama [56] is not essential. If electrons are supplied through or around the oxide film, a charged species can form. Otherwise, oxygen molecules are the mobile species at temperatures around 1000°C and 1 atm oxygen pressure.

Modifications of the oxidation model of Deal and Grove have been proposed. Interest has centered on the anomalously high initial oxidation rate which is found prior to the linear-parabolic region of dry oxidation. Blanc [57] and Ghez and van der Meulen [58] put forward a model to explain this initial high rate. Molecular oxygen transport through the oxide is assumed, but the reaction of atomic rather than molecular oxygen is postulated to occur at the oxide–silicon interface, a model subsequently considered by Tiller [47]. Blanc's model predicts a $P^{1/2}$ dependence of the linear rate constant. A $P^{1/2}$ dependence has been observed for oxidation up to an oxide thickness of 500 Å [59], but the kinetics were undefined. Blanc fits the data of Irene et al. [60,61] to his equation, finding good agreement.

Irene and van der Meulen [60] interpret their own data for dry oxidation in terms of the linear-parabolic model in an attempt to explain the quantitative lack of agreement in the literature. They found an initial oxidation regime up to a thickness of 350 Å which did not fit this model, although data between 350 and 1100 Å did fit. They concluded that the variation in rate constants reported in the literature originates from several factors rather than just one. Water concentrations greater than 25 ppm were found to be significant [62]. Nakayama and Collins [63] had proposed that hydrogen from water could "catalyze" silicon oxidation if hydroxyl groups were important as diffusing species. Also considered important were the form in which the linear-parabolic rate equation was expressed, the use of data outside the best-fit regime, and an insufficiency of data. These all contributed to differing values of the rate constants. Based on their activation energy for the linear rate constant, they favored the breaking of an O–O bond rather than a Si–Si bond. In a similar departure from Deal and Grove, they suggested that a 2.3-eV value of activation energy for parabolic growth is too large to correlate with molecular oxygen diffusion.

Van der Meulen [64] studied the rate of oxide growth for thicknesses less than 300 Å. He found that, for temperatures of 700–1000°C and oxygen pressures of 0.01 to 1.0 atm in a N_2–O_2 mixture at 1 atm total pressure, oxide thickness differs on (100) and (111) oriented wafers. A different pressure dependence for the two orientations was interpreted in terms of competing reactions involving atomic and molecular oxygen.

$$\text{Si–Si} + O_2 \rightarrow \text{Si–O–Si} + O \tag{22}$$

$$\text{Si–Si} + O \rightarrow \text{Si–O–Si} \tag{23}$$

$$O_2 \rightleftarrows 2O \tag{24}$$

Equation 22 was taken to be dominant at high temperatures while Equation 24 was assumed to dominate at low temperatures. Equation 23 was essential in both cases. Kinetic expressions for oxide growth are derived by Ghez and van der Meulen [58] based on the above reactions. Irene [65] also offers support for the parallel rate processes in terms of Arrhenius plots.

Revesz, in a series of papers [33], argues for the existence of channels in dry oxide, as discussed in Chapter 3. The channels are postulated to form perpendicularly to the surface of the silicon through a localized ordering of the silica structure. Such structural channels should be small, less than 50 Å in diameter. Irene [66] postulated the existence of pores less than 50 Å in diameter which both change the initial oxidation kinetics and lower the dielectric breakdown strength of the oxide. Revesz [67] does not agree with Irene concerning this latter point. Gibson and Dong [68] report the observation by transmission electron microscopy of 10-Å pores typically 100 Å apart in dry SiO_2 films. Silica films grown under wet conditions fail to show such pores. Srivastava and Wagner [69] also report the existence of micropores with 5–8 Å diameter in thin dry oxide. The micropores were randomly distributed. It is possible that, in the present discussion, channels and pores represent similar entities.

Some ordering of a crystalline phase of SiO_2 was observed in the thin oxide [69]. Agius et al. [70] also observed a structural order in SiO_2 films ~11 Å thick.

Channels have been postulated to disappear as the oxide thickens under dry conditions or fail to form in the presence of water [33,71]. The data of Rosencher et al. [27] have been fitted to a linear-parabolic rate law by Revesz and Schaeffer [8] and they find O_2 permeation rates larger than those found in bulk vitreous silica. They point out that this result is consistent with a model that includes channels during the early stages of dry oxidation. Thus, x_i can be xero for wet oxidation and finite for dry.

9.3.2. Effect of Electric Field

Field-assisted oxidation of silicon is another process which needs discussion. Mackintosh and Plattner [72], using noble gas markers, have shown that anionic species are responsible for anodic oxidation of silicon in a variety of electrolytes. Jorgensen [73] carried out solid-state anodization of silicon at 700–850°C and showed that anions are involved. Four electrons were transferred for each SiO_2 formed at the anodic interface. These results caused some controversy over the role of the platinum [74] and as to whether anodization is relevant to thermal oxidation [75]. Attempts to answer this latter question have continued through work such as that of Laverty and Ryan [76] and Mills and Kroger [77]. Raleigh's

[75] point is that the external connection used by Jorgensen provides electrons which during oxidation must be supplied through the oxide. Only in the case of rate limitation by ion movement would anodization and oxidation be equivalent. This appears to occur quite often. See Chapter 2.

Schaeffer [28] has put forth a correlation which may be useful in resolving the question of neutral versus charged species as the oxidant for silicon. His model of oxygen molecular movement in SiO_2 is based on an interaction between dissolved oxygen and network oxygen as found during isotopic exchange measurements [78,79,80,81,82]. In order that $^{18}O_2$ can exchange with Si–^{16}O–Si, an ^{16}O–Si bond must break. The "interstitially" dissolved O_2 may form an ozonelike complex with an oxygen atom of the silica network [83] or perhaps a peroxide radical [53]. Rotation or breakup of the "interstitial" defect then provides for isotopic exchange as well as oxidant movement during oxidation. The process is analogous to an interstitialcy mechanism. An analogy can also be drawn with vacancy diffusion in crystals where the permeability of vacancies equals that of atoms. In the case of vitreous silica, the permeability of molecular oxygen is assumed to be equal to that of network oxygen such that the diffusing oxygen must interact with network oxygen.

Permeability equals diffusivity D times concentration C. The resulting correlation is

$$D_O \cdot C_O = D_{O_2} \cdot C_{O_2} \tag{25}$$

where C_O is the concentration of network-bridging oxygens and C_{O_2} is the solubility of molecular O_2 in SiO_2. The concentration C_{O_2} depends directly upon oxygen pressure while D_{O_2} and C_O do not. It has been found in isotopic exchange studies that D_O also depends directly on oxygen pressure. Thus, both sides of Equation 25 are pressure dependent. In addition, both sides have similar activation energies,

$$Q'(D_{O_2}) = Q'(D_O) = 1.04 \text{ eV} \tag{26}$$

while concentrations are independent of temperature.

A caution must be inserted here. The presence of water during isotopic exchange experiments can influence the exchange rate of oxygen [42]. Rosencher et al. [27] showed, using nuclear microanalysis, that ^{18}O does not exchange with thermally grown Si–$^{16}O_2$ films on silicon except in a thin surface layer. Rochet et al. [84] have confirmed that this surface exchange is due to oxygen rather than water. Revesz and Schaeffer [8] explained the limited amount of exchange in the rest of the oxide in terms of a change in oxide structure, that is, the existence of channels.

It is well known that water, as it permeates into SiO_2, reacts with the network to form SiOH [85,86,87]. Hydrogen may also be involved in this process [88]. See Equation 4. As a result, a modified relationship has been constructed [28] for the simultaneous diffusion of H_2O in SiO_2 and reaction between H_2O and SiO_2. Thus, the correlation

$$2D_{H_2O} \cdot C_{H_2O} = D_{SiOH} \cdot C_{SiOH} \tag{27}$$

results, where D and C have the same meanings as above, H_2O refers to molecular water, and SiOH to a product of the reaction in Equation 3. The above correlations require further experimental verification, but if they survive such a test, it becomes possible to speak in terms of molecular and/or network constituents when discussing the oxidation of silicon.

These ideas were extended by Revesz and Schaeffer [8] who differentiate solubility sites from network sites in SiO_2. They also conclude that a structural ordering accompanies the diffusion of network oxygen.

A model for the growth of both thick and thin SiO_2 on Si has been presented by Lu and Cheng [89]. They postulate that an internal electric field affects the oxidizing species O_2^-.

9.3.3. Defects in SiO_2

The existence of defects in vitreous oxides [90], as pointed out in Chapter 4, is one key to unraveling the dilemma of silicon oxidation by neutral versus charged species. The chief defects in vitreous silica consist of a nonbridging oxygen and trivalent silicon. They simultaneously form from the breaking of an Si–O–Si bridging bond. The resulting nonbridging or dangling oxygen, $-Si-O^{\delta-}$, carries a partial negative charge when it retains the electron of the adjoining silicon and is referred to as D^- by Mott [91]. The trivalent silicon $-Si^{\delta+}$ carries a partial positive charge and is designated as D^+. Lucovsky [92] further discusses defects in vitreous silica. He postulates three-coordinated oxygen, designated as C_3^+ and calls the three-coordinated silicon T_3^+. Reactions of defects are discussed using this nomenclature. See Chapter 4 for further discussion.

The concentration of broken bonds in a network may be estimated by analogy with Schottky defect formation.

$$n_d = n_c^0 \exp\left(-\frac{\Delta H_d^0}{kT}\right) \tag{28}$$

where n_d = volume concentration of defects
n_c^0 = concentration of bonds available for breaking
ΔH_d^0 = formation enthalpy for the defect pair.

Dignam [93] proposes that a linear relationship may exist between the experimental activation energy for anodic film growth and the calculated enthalpy of formation ΔH_d^0 for network defect pairs in the case of oxides which are vitreous. Interpolating his relationship and extending it to even-valent oxides, one obtains a value of $\Delta H_d^0 = 3.0$ eV for SiO_2. One can then calculate from Equation 28, taking $n_c^0 = 2 \times 10^{23}$ cm^{-3}, a value of n_d equal to 3×10^{11} cm^{-3} at 1000°C. This value of n_d is much lower than the interface trap densities reported by Caplan et al. [94] and Poindexter et al. [95] or the solubilities of O_2 or H_2O in SiO_2 as given in Table 9.1. There are no strong indications at this time that the solubility sites are related to the network defects. However, the latter may possibly provide the charged entities required for field-driven oxidation.

The reduction of oxygen pressure by several orders of magnitude, as in the work of Kamigaki and Itoh [15], results in high-temperature oxidation kinetics which fit the Cabrera–Mott rate law. This implies ion movement under the influence of an electric field.

An explanation [96] for the change from linear-parabolic kinetics at higher pressures to inverse logarithmic kinetics at lower pressures can be proposed as follows. However, it must be kept in mind that the ideas have yet to be tested thoroughly and are only presented here in view of the above discussion.

The quantity of dissolved oxygen at lower pressures is insufficient to support molecular movement and the interfacial reactions as given by Equations 22–24. Instead, electronic transport builds up a spontaneous voltage across the thin oxide. The field thus created aids ion movement, leading to Cabrera–Mott kinetics.

Electronic transport by tunneling or thermionic emission can provide electrons to fill oxygen traps on the oxide surface [97]. See Chapter 4. Silica films are known to support electronic conduction during anodization. Current in the liquid-phase process at 25°C is 99% electronic [98] while solid-state anodization [73] was found to have an electronic transference number of 0.6. The charge separation due to the trapped electrons sets up an electric field which lowers the activation energy for ion movement. It is proposed that the ions in this case are related to nonbridging oxygen defects as in the model of Dignam [93]. Through movement of such defects, the oxide film would continue to grow. In essence, at high temperatures there would be a gradual change from molecular mass transport at higher pressures to ionic mass transport at lower pressures.

Support for this model of silicon oxidation is found in the work of Rochet et al. [84]. They used an ^{18}O tracer to study the oxidation mechanism at 930°C. Oxidation in natural $^{16}O_2$ was followed by growth in $^{18}O_2$. The oxide was then profiled using nuclear microanalysis. They found ^{18}O at both the Si/SiO_2 and SiO_2/O_2 interfaces. Most of the ^{18}O was located at the Si/SiO_2 interface for films ≧ 200 Å thick, but when film thickness approached 50 Å, 50–80% of the ^{18}O was found at the SiO_2/O_2 interface. This result could be explained by molecular oxygen incorporation at the Si/SiO_2 interface and oxygen ion incorporation at the SiO_2/O_2 interface.

Competition between molecular and ionic movement may be present in an anodization experiment such as Jorgensen's [73]. Here, a porous platinum electrode was deposited over the oxide so that resistance changes in the oxide could be measured and/or an electric field applied. The presence of the platinum electrode also provided a means for dissociating oxygen and ionizing the resulting atoms. This process is observed in zirconia-based oxygen sensors [99] which provide a model for the Si–SiO_2–Pt system. In addition to surface ions, O_2 molecules saturate the SiO_2. Jorgensen found that a field of $\sim 10^4$ V cm^{-1} at 850°C increased or decreased the rate of oxidation for a negative or positive polarity of the Pt electrode, respectively. Oxide growth was found to stop at an inhibiting potential of 1.62 V on the Pt electrode. This value was consistent with the free energy change for the formation of SiO_2. A higher voltage resulted in the reduction of silica and epitaxial growth of silicon crystallites at the oxide–silicon interface [100].

9.3.4. Summary of Oxidation Versus Anodization

Thermal energy is insufficient to account for continued oxidation according to the Deal–Grove model at low temperatures, say below 500°C, and 1 atm pressure. Instead, the self-induced field of the Cabrera–Mott theory can be invoked and found to fit the experimental results. See Chapter 8 for results on silicon. Ionic movement dominates as in anodization. See Chapter 4. The reaction is interface controlled. The anionic species, perhaps a vacancy generated at the Si/SiO_2 interface, once in the oxide, is drawn rapidly to the opposite interface where it reacts.

We find then that the present models for the formation of vitreous SiO_2 films from crystalline silicon depend upon one or more of the following observations. Vitreous silica may contain both charged defects and dissolved oxygen molecules in its open network structure. Mass transport during oxidation occurs as a result of the movement of an oxygen-based species. Molecules appear to dominate this transport at high temperatures and pressures. However, low temperatures and/or low oxygen pressures along with the presence of an electron source can result in the movement of charged species. The presence of micropores and the viscoelastic properties of SiO_2 modify these processes. As a result, a universal model of silicon oxidation is still evolving.

9.4. NOVEL METHODS OF FORMING SiO_2 FILMS

Adsorption of oxygen on silicon and the growth of thin SiO_2 films have been treated in Chapter 8. However, several new methods of oxidation based on gas-phase anodization, ion implantation, and lasers have evolved in the search for lower temperature processing of integrated circuits. Chemical and physical vapor deposition of oxides have been omitted from this discussion.

9.4.1. Gas-Phase Anodization

Liquid-phase anodization of silicon [101] is possible, but the efficiency is poor. As a result, gas-phase anodization [102] has been developed as a low-temperature, dry process. Reviews by Ray [103] and by Sugano [104] define the conditions of growth using a DC or high-frequency electrical discharge. At 600°C, a fast growth rate of 10^4 Å hr^{-1} can be achieved, using an oxygen pressure of 0.2 Torr and 420-kHz discharge frequency (1 kW). A low lateral growth rate is experienced under these conditions, leading to better geometries at the oxide edge. The DC voltage applied between the silicon anode and silicon cathode is ~200 V for a 1 μm thick film. The dielectric and chemical properties of the oxide formed are the same as those formed by thermal oxidation. However, the density of interface-trapped charge at the Si/SiO_2 interface is an order of magnitude larger. The number of electronically active interface states can be reduced to 10^{10} cm^{-2} eV^{-1} by annealing in forming gas or chlorine. Pulfrey et al. [105] report on progressive buildup of positive charge in the oxide. This could be dissipated by a low-temperature anneal.

Evidence is presented [104] which indicates that both cationic and anionic species move through the oxide. This result differs from that of Mackintosh and Plattner [72] who show that only anion movement occurs during the liquid-phase anodization of silicon. Siejka and Perriere [106] also report that oxygen ions are the moving species during plasma oxidation of silicon.

Kinetics of SiO_2 growth in a plasma have been interpreted in terms of the Cabrera–Mott rate law [106,107]. MOS tunneling devices based on plasma-formed silica films are discussed by Atanassova et al. [108].

9.4.2. Ion Implantation

Watanabe and Tooi [109] reported the formation of SiO_2 films by oxygen ion bombardment of silicon. Properties of the films matched those of thermally grown oxide, except for a voltage-independent capacitance. This was caused by radiation damage for doses greater than 10^{14} ions cm^{-2} at 60 keV.

Dylewski and Joshi [110] established that 10^{18} ions cm^{-2} and annealing at 550–800°C were needed to form an SiO_2 layer using 30-keV ions. Kirov et al. [111] extended the annealing range to 700–1000°C. A voltage-dependent capacitance was achieved using 10^{18} ions cm^{-2}, annealing in N_2 at 1000°C and then in H_2 at 450°C. Chiang et al. [112] report on the laser annealing of oxygen ion-implanted silicon. A granular oxide structure was observed in TEM.

9.4.3. Light-Induced Oxidation

Oxide formation was observed by Liu et al. [113] during the UV laser melting of a single crystal silicon surface to produce amorphous silicon. Substitution of Ar for O_2 showed conclusively that a gas-phase reaction was involved. Oren and Ghandhi [114] had previously showed that ultraviolet radiation could enhance the oxidation of silicon at high temperatures. Schafer and Lyon [115] subsequently reported a wavelength dependence for the phenomenon. Visible radiation enhanced the dry thermal oxidation rate ~40% while UV (350 nm) increased it ~60% for the range 770–900°C. In the case of wet oxidation, both wavelength regions increased the rate ~40%. They concluded that some component of dry oxidation, not present or important in wet oxidation, is affected by high-energy photons. A threshold energy of 3.0–3.5 eV appeared to exist. While part of the effect of laser radiation is thermal, Young and Tiller [116] have confirmed that a photonic effect certainly exists when using dry oxygen. For photon energies in the range 2.4–2.7 eV, the enhanced oxidation rate is linearly proportional to the photon flux density. Si (100) showed greater enhancement than Si (111), opposite to the expectations of a purely thermal effect. They point out that high power densities create more heating of the silicon which eventually washes out the photonic effect as found by Oren and Ghandhi [114]. The use of CO_2 lasers to induce oxidation of silicon is also effective [117,118]. Here, a thermal effect is assumed to dominate.

9.5. DEFECTS AT THE SILICON–SILICON DIOXIDE INTERFACE

The beauty of the Si–SiO_2 couple is the availability of single crystal silicon as a probe of charged states at the semiconductor–oxide interface. These states, which can lead to device instabilities, are related to the process of oxidation and thus reveal details which cannot be observed in metal–oxide systems. This topic bears directly on silicon device technology which is well covered in a book by Nicollian and Brews [4]. Hence, mainly

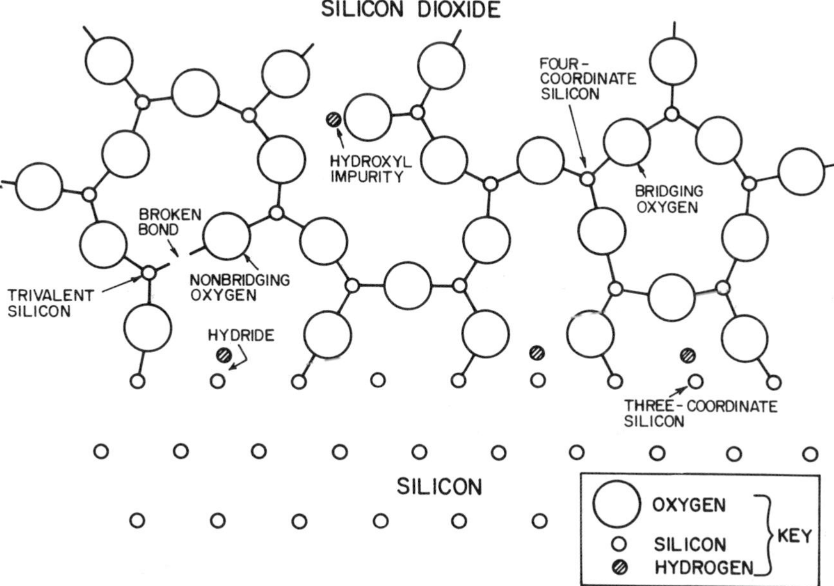

Figure 9.10. Cross section of the Si/SiO_2 interface showing major defects. The silicon atoms are shown as three- rather than four-coordinate because of the two-dimensional nature of the drawing.

those aspects pertaining to vitreous oxide structure will be emphasized here.

Following the early discovery by Atalla et al. [119] that thermally grown SiO_2 stabilizes the surface of Si, Kooi [120,121,122] presented an $Si–SiO_2$ model which included defects in SiO_2 near the Si/SiO_2 interface as well as defects created by structural mismatch between the crystalline silicon and vitreous oxide. These are discussed below. The model has been modified and added to over the years, but it remains the basis for interpreting the $Si–SiO_2$ system. The most significant change has been experimental; more sensitive detection of defects and a more exact measurement of interface width yielding the current value of <10 Å. See Chapter 6.

A schematic drawing of the $Si–SiO_2$ couple, including impurities and defects, is shown in Figure 9.10. Dangling bonds are found because of incomplete saturation of the silicon interface with oxygen, as well as broken silicon–oxygen bonds within the oxide. Impurities such as hydrogen act to saturate these broken bonds. Halogens can be made to act in a similar fashion although they must be built into the oxide rather than diffused in later as in the case of hydrogen.

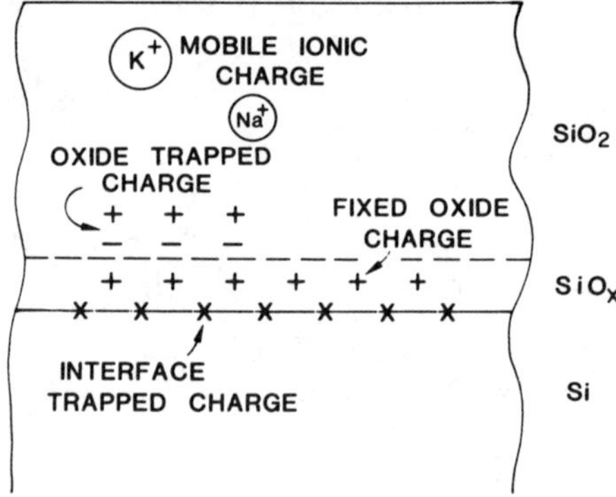

Figure 9.11. Names and location of major charges found in thermally oxidized silicon. After Deal [123]. Reprinted by permission of the publisher, The Electrochemical Society, Inc.

Deal [123] has published a standardized terminology for charges associated with oxidized silicon. The principal charges in Si–SiO_2 are shown schematically in Figure 9.11. These are discussed below where Q is the charge and N is the number of charges per unit area.

9.5.1. Fixed Oxide Charge

Fixed oxide charge (Q_f, N_f) is positive and caused by defects in the oxide, for example, the trivalent silicon shown in Figure 9.10. A higher density of charge is found for SiO_2 thinner than 500 Å [124]. The defects which account for fixed oxide charge appear to be located outside the electron tunneling distance, that is, $\geqq$ 25 Å from the silicon surface, since there is no electrical communication between the defects and the silicon. Their existence is very process dependent, varying with temperature, environment, annealing schedule, and silicon orientation.

9.5.2. Oxide-Trapped Charge

Oxide-trapped charge (Q_{ot}, N_{ot}) in the oxide can be positive or negative due to trapped holes or electrons, respectively. These charges are produced by ionizing radiation, avalanche injection, or similar processes. Thus, the traps can be in electrical communication with the silicon surface. Annealing at less than 500°C generally reduces the charge concentration, leaving neutral traps. Hydrogen plays a role in this process [125,126]. Mott et al. [53,127] discuss the existence of positive and negative charges resulting from dangling bonds. These charges are invoked to explain the Anderson localization of holes and electrons at the Si/SiO_2 interface. They are discussed in terms of D^0, D^+, and D^- [91].

9.5.3. Mobile Ionic Charge

Mobile ionic charge (Q_m, N_m) at room temperature is attributed to impurity ions such as Li^+, Na^+, K^+, and possibly H^+. Negative ions such as Cl^- can compensate the positive charge of such impurities, but can themselves become mobile at temperatures greater than 500°C. Divalent cations may also become mobile at higher temperatures, as found in bulk glasses [128] where divalent ions are utilized to block the movement of alkali ions. The concentration of mobile sodium can be reduced during processing by gettering with chlorine during dry oxidation [129,130,131]. A similar complexing of iron by fluorine in glass is discussed by Weyl [132]. A two-step HCl process, oxidation followed by an anneal, reduces the number of oxide defects [133,134], but too much chlorine can result in film detachment [130]. RF plasma cleaning [135] or a postanneal in inert gas followed by a short reoxidation [136] is also effective in limiting the number of defects in thin SiO_2 films.

The mobile ionic charge concentration is found to be higher at the oxide interface than in the bulk. The surface concentration appears to be limited by the reducing power of silicon [121]. For instance, sodium ions may be converted to neutral metal at the Si/SiO_2 interface. The sodium atoms can then diffuse back into the oxide.

Sodium can also be gettered from the oxide by reacting it with phosphorus oxide at the oxide–gas interface. The layer of phosphosilicate glass thus formed has a much greater affinity for alkali than the SiO_2 itself. See Chapter 3 for a discussion of the need for compensation when a pentavalent phosphorus substitutes for a tetravalent silicon in a network structure. Vitreous oxide structure would predict that boron or aluminum oxide should also getter sodium when assuming tetrahedral coordination in the silica network.

A useful employment of mobile ionic charge is in a chemically sensitive MOS element [137].

The concentration of Q_f, Q_m, and Q_{ot} can be determined from a measurement of high-frequency capacitance versus surface potential (C–V technique). Low-frequency quasistatic C–V or I–V analysis or some other electronic technique [4,138], for example, deep-level transient spectroscopy (DLTS) [139,140], is required to measure Q_{it} (see below, and Figure 9.12). C–V measurements as a function of temperature provide information about mobile ionic charge. Polarization of the phosphosilicate glass used to getter sodium can also be detected [141]. Polarization can be eliminated by stripping the phosphosilicate layer with dilute hydrofluoric acid after completing the gettering cycle. A polarization phenom enon is also found in lead silicate films [142] used as sodium getters.

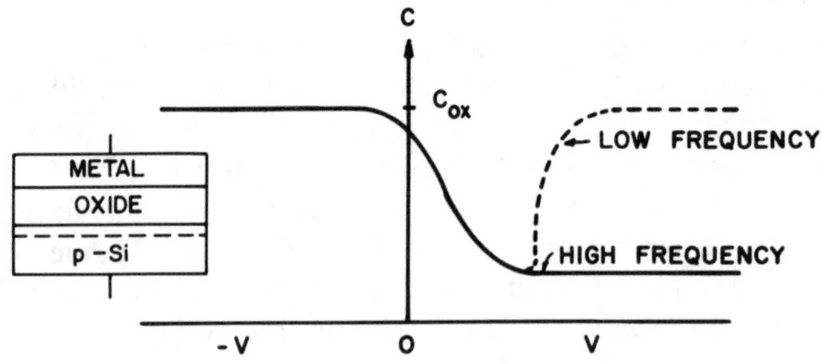

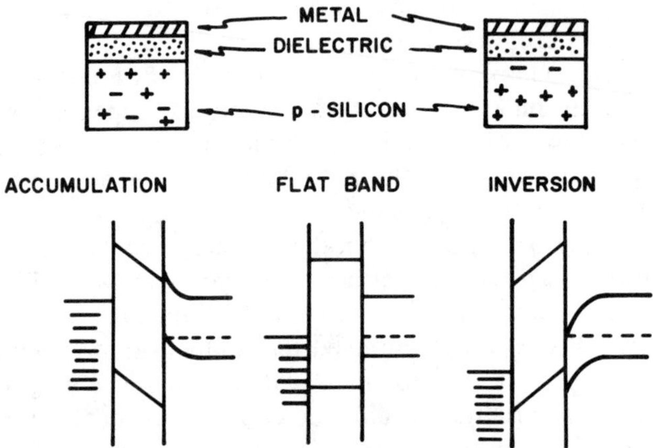

Figure 9.12. Capacitance–voltage curve at low and high frequency for a metal-oxide-semiconductor diode as shown schematically to the left. The state of charge of the silicon surface is shown below as a function of voltage on the metal. See Reference 4 for further discussion.

9.5.4. Interface-Trapped Charge

Interface-trapped charge (Q_{it}, N_{it}) can be positive or negative. Thus, the threefold coordinate silicon at the Si/SiO_2 interface (Figure 9.10) can act as either a donor or an acceptor, depending on the location of the Fermi level in the silicon. Sources of defects in addition to those created during the oxidation process are impurities located at the interface or broken bonds caused by ionizing radiation. The charges are in easy electrical communication with the silicon as contrasted with fixed or mobile charge. As a result, interface traps are charged or discharged by changes or fluctuations in the surface potential. Low-temperature (450°C) hydrogen annealing can neutralize the charge [143], leading to Si–H or Si–OH bonds as shown at the interface in Figure 9.10. Other names used in the past to describe Q_{it} are surface states, fast states, or interface states. The dependence of interface-trapped charge on surface potential is expressed by interface trap level density D_{it} (number cm^{-2} eV^{-1}) which equals $N_{it}/\mathscr{E}_F$ where $\mathscr{E}_F$ is the Fermi level in the silicon and $N_{it} = \pm Q_{it}/q$. Figure 9.10 attempts to summarize the Si/SiO_2 interfacial structure. The chemical properties are dominated by the vitreous nature of the oxide structure. Electrical properties, on the other hand, are controlled by defects such as dangling bonds and impurities.

A relationship between interface-trapped charge and fixed-oxide charge has been proposed by Razouk and Deal [144]. However, it has not been confirmed experimentally. An ESR center located at the Si/SiO_2 interface and identified with dangling bonds at the (111) silicon surface (see Figure 9.10) correlated with Q_{it}, but not with Q_f [94]. Q_f also did not correlate with E' centers (trivalent silicon) in the oxide. On (100) silicon wafers, the ESR signal had two components, the $\cdot$ $Si{\equiv}Si_3$ discussed above and $\cdot$ $Si{\equiv}Si_2O$. The proportionality between the ESR center and Q_{it} was maintained, but both were three times smaller than on (111) silicon [95].

Hahn [145] correlated atomic structure of the Si/SiO_2 interface with electrical properties. He claims that both interface states and fixed-oxide charge have at least one common physical origin, the atomic roughness.

Nitridation of thin oxide on silicon has been studied between 900 and 1150°C at 0.7 to 3675 Torr NH_3 as a method for improving the gate oxide of metal-insulator-semiconductor (MIS) devices. The nitrogen released by ammonia after 5 hr at 1100°C and 735 Torr appears to be incorporated throughout the oxide [146]. For times under 2 hr at 1 Torr NH_3 and ~1000°C, nitrogen is found only at the interfaces [147]. The nitrogen affects the density of interface-trapped charge, increasing it by more than a factor of 2.

9.6. EFFECT OF DOPANTS

Deliberate introduction of dopant impurities into silicon is a key part of IC device technology. Some dopants are added during silicon crystal growth to uniformly change electrical properties while others are introduced in a controlled manner to change properties in localized areas. In both cases, interaction with a surface oxide affects the final dopant concentration.

The rate of oxidation of doped silicon is unaffected by small concentrations of boron or phosphorus, in the 10^{16} to 10^{19} cm^{-3} range [4]. However, effects are observed when the dopant concentration becomes comparable to the silicon atom concentration, in the oxide itself for the parabolic regime of oxide growth and on the silicon surface for the linear regime. These effects become significant for dopant concentrations of 10^{20} cm^{-3} or greater. An electronic model has been put forward by Ho and Plummer [148] to explain enhanced oxidation rates of heavily doped silicon. An increase in the equilibrium concentration of point defects in the silicon substrate is proposed.

Redistribution of the dopant due to segregation between silicon and SiO_2 occurs during oxidation. Boron segregates to the oxide and increases the diffusion controlled parabolic rate by weakening the oxide structure. The presence of water decreases the effect by decreasing the boron concentration in the oxide. Phase separation within the oxide has been noted [26]. Phosphorus, on the other hand, segregates to the silicon. This increases the linear rate which is interface controlled.

Doping of the silicon surface is carried out using a mask which blocks dopant diffusion, thus limiting the areas where conductivity change takes place. A silica mask is effective for B or P until the SiO_2 is converted to a phospho- or borosilicate glass. Such glasses can be used directly as a dopant source. A useful oxide mask has a diffusion coefficient for the impurity which is lower than that for the impurity in silicon. Thus, SiO_2 is used in masking P, B, As, and Sb, but not gallium. Frischat [128] has collected the diffusion data for B, Ga, P, As, Sb, Au, Ni, Na, Li, K, Ag, Ca, "water," and oxygen in bulk fused silica or quartz. There is a great deal of scatter in the data. Grove [149] summarizes data for diffusion in both Si and SiO_2.

9.7. CHEMICAL ETCHING

Evaluation of vitreous SiO_2 films involves the measurement of stoichiometry as discussed in Chapter 8 and electrical properties as described above. Chemical etching also reveals a great deal about the structure and composition of such films. For instance, a comparison between etch rates in buffered hydrofluoric acid and P-etch (5 parts HNO_3 to 3 parts acetic acid to 3 parts HF) shows that silicon dioxide etches faster in the former while borosilicates etch faster in the latter [150]. The etch rate combined with optical properties provides a convenient assessment of density, excess silicon, bond strain, and Al_2O_3 or PbO content [151]. Characterization by Raman spectroscopy is a relatively new method of determining the structure and composition of vitreous oxides [152,153,154].

9.8. CONCLUSIONS

Vitreous silica is uniquely suited to the needs of both the integrated circuit and optical waveguide industries. The flexibility of its network structure allows it to conform on an atomic level with the silicon crystal lattice. As a result, most bonds are completed within the oxide and at the Si/SiO_2 interface so that the number of optical and electronic defects is minimized. In addition, some impurities can be incorporated into the vitreous oxide structure, rendering them electrically inactive. The network structure lends itself to forming smooth surfaces free of grain boundaries and dislocations. This is important in integrated circuits for the fabrication of metal interconnects.

The rate of conversion of silicon into silica by thermal oxidation is described by the Deal–Grove equation. Extensive modification of this model is occurring as experimental conditions of temperature, pressure, and applied field are extended.

The properties of vitreous silica include chemical stability, although it can be etched if necessary. The oxide serves both as a mask against dopants such as P or B and as a source of the same dopants if it is presaturated with them.

Both ionic and molecular movement can occur in vitreous silica. Diffusion of impurity ions such as sodium is well documented. The mechanism of oxygen self-diffusion is still a subject of debate. The role of hydrogen in breaking up the oxide network as well as destroying defects at the Si/SiO_2 interface is a key part of the chemistry of vitreous silica.

REFERENCES

1. J. S. Williams and C. E. Christodoulides, *Nucl. Instrum. Methods* **182/183** (1981), 667.
2. J. Ruzyłło, I. Shiota, N. Miyamoto, and J-i. Nishizawa, *J. Electrochem. Soc.* **123** (1976), 26.
3. A. K. Zakzouk, R. A. Stuart, and W. Eccleston, *ibid.* **123** (1976), 1551.
4. E. H. Nicollian and J. R. Brews, *MOS (Metal-Oxide-Semiconductor) Physics and Technology*, John Wiley & Sons, New York, 1982.
5. W. H. Dumbaugh and P. C. Schultz, *Encyclopedia of Chemical Technology*, 2nd ed., Vol. 18, John Wiley & Sons, New York, 1969, p. 73.
6. A. G. Revesz, *J. Non-Cryst. Solids* **11** (1973), 309.
7. R. H. Doremus, *J. Phys. Chem.* **80** (1976), 1773.
8. A. G. Revesz and H. A. Schaeffer, *J. Electrochem. Soc.* **129** (1982), 357.
9. R. H. Doremus, *Glass Science*, John Wiley & Sons, New York, 1973, p. 121ff.
10. T. L. Barr, *Applic. Surface Sci.* **15** (1983), 1.
11. F. S. Grunthaner and J. Maserjian, in *The Physics of SiO_2 and its Interfaces*, S. T. Pantelides, Ed., Pergamon Press, New York, 1978, p. 389.
12. F. J. Grunthaner, B. F. Lewis, and J. Maserjian, *J. Vac. Sci. Tech.* **20** (1982), 747.
13. F. W. Smith and G. Ghidini, *J. Electrochem. Soc.* **129** (1982), 1300.
14. J. Derrien and M. Commandré, *Surface Sci.* **118** (1982), 32.
15. Y. Kamigaki and Y. Itoh, *J. Appl. Phys.* **48** (1977), 2891.
16. J. R. Ligenza, *J. Electrochem. Soc.* **109** (1962), 73.
17. H. Rawson, *Inorganic Glass-Forming Systems*, Academic Press, New York, 1967, pp. 48–61.

18. F. E. Wagstaff and K. J. Richards, *J. Am. Ceram. Soc.* **48** (1965), 382.
19. B. E. Deal and A. S. Grove, *J. Appl. Phys.* **36** (1965), 3770.
20. A. G. Revesz and R. J. Evans, *J. Phys. Chem. Solids* **30** (1969), 551.
21. G. A. Haas and H. F. Gray, *J. Appl. Phys.* **46** (1975), 3885.
22. M. A. Hopper, R. A. Clarke, and L. Young, *J. Electrochem. Soc.* **122** (1975), 1216.
23. Y. Ota and S. R. Butler, *ibid.* **121** (1974), 1107.
24. R. R. Razouk, L. N. Lie, and B. E. Deal, *ibid.* **128** (1981), 2214.
25. L. N. Lie, R. R. Razouk, and B. E. Deal, *ibid.,* **129** (1982), 2828.
26. E. A. Irene and D. W. Dong, *ibid.* **125** (1978), 1146.
27. E. Rosencher, A. Straboni, S. Rigo, and G. Amsel, *Appl. Phys. Lett.* **34** (1979), 254.
28. H. A. Schaeffer, *J. Non-Cryst. Solids* **38 and 39** (1980), 545.
29. S. S. Cristy and J. B. Condon, *J. Electrochem. Soc.* **128** (1981), 2170.
30. D. J. Breed and R. H. Doremus, *J. Phys. Chem.* **80** (1976), 2471.
31. U. R. Evans, *The Corrosion and Oxidation of Metals*, St. Martins Press, New York, 1960, p. 826.
32. R. J. Zeto, E. Hryckowian, C. D. Bosco, G. J. Iafrate, R. W. Brower, and C. G. Thornton, Abstract No. 206, p. 504, The Electrochemical Society Extended Abstracts, Vol. 74-2, New York, Oct. 13–17, 1974.
33. A. G. Revesz, *Phys. Stat. Sol.* **A57** (1980), 235, 657; **A58** (1980), 107.
34. S. I. Raider and L. E. Forget, *J. Electrochem, Soc.* **127** (1980), 1783.
35. L. Pauling, *The Nature of the Chemical Bond*, 3rd ed., Cornell University Press, Ithaca, NY, 1960.
36. F. J. Norton, *Nature* **191** (1961), 701.
37. E. P. EerNisse, *Appl. Phys. Lett.* **30** (1977), 290; **35** (1979), 8.
38. E. A. Irene, D. W. Dong, and R. J. Zeto, *J. Electrochem. Soc.* **127** (1980), 396.
39. E. A. Irene, *ibid.* **129** (1982), 413.
40. E. A. Irene, *J. Appl. Phys.* **54** (1983), 5416.
41. A. J. Moulson and J. P. Roberts, *Trans. Faraday Soc.* **57** (1961), 1208.
42. S. Rigo, F. Rochet, B. Agius, and A. Straboni, *J. Electrochem. Soc.* **129** (1982), 867.
43. K. H. Beckmann and N. J. Harrick, *ibid.* **118** (1971), 614.
44. P. J. Burkhardt, *ibid.* **114** (1967), 196.
45. J. E. Shelby, *J. Appl. Phys.* **48** (1977), 3387.
46. I. Abbati, G. Rossi, L. Calliari, L. Braicovich, I. Lindau, and W. E. Spicer, *J. Vac. Sci. Technol.* **21** (1982), 409.
47. W. A. Tiller, *J. Electrochem. Soc.* **127** (1980), 619, 625; **128** (1981), 689; **130** (1983), 501.
48. R. H. Doremus, *Thin Solid Films* **122** (1984), 191.
49. E. A. Taft, *J. Electrochem. Soc.* **125** (1978), 968.
50. M. Hirayama, H. Miyoshi, N. Tsubouchi, and H. Abe, *IEEE Trans.* **ED-29** (1982), 503.
51. M. Hirayama, H. Miyoshi, N. Tsubouchi, and H. Abe, *J. Electron. Mater.* **11** (1982), 919.
52. E. A. Irene, E. Tierney, and J. Angilello, *J. Electrochem. Soc.* **129** (1982), 2594.
53. N. Mott, *Proc. Roy. Soc.* (*London*) **A376** (1981), 207.
54. N. F. Mott, *Phil. Mag.* **A45** (1982), 323.
55. J. A. Costello and R. E. Tressler, *J. Electrochem. Soc.* **131** (1984), 1944.
56. F. C. Collins and T. Nakayama, *ibid.* **114** (1967), 167.
57. J. Blanc, *Appl. Phys. Lett.* **33** (1978), 424.
58. R. Ghez and Y. J. van der Meulen, *J. Electrochem. Soc.* **119** (1972), 1100.
59. T. Smith and A. J. Carlan, *J. Appl. Phys.* **43** (1972), 2455.
60. E. A. Irene and Y. J. van der Meulen, *J. Electrochem. Soc.* **123** (1976), 1380.

61. E. A. Irene and R. Ghez, *ibid.* **124** (1977), 1757.
62. E. A. Irene, *ibid.* **121** (1974), 1613.
63. T. Nakayama and F. C. Collins, *ibid.* **113** (1966), 706.
64. Y. J. van der Meulen, *ibid.* **119** (1972), 530.
65. E. A. Irene, *Appl. Phys. Lett.* **40** (1982), 74.
66. E. A. Irene, *J. Electrochem. Soc.* **125** (1978), 1708.
67. A. G. Revesz, *ibid.* **126** (1979), 502.
68. J. M. Gibson and D. W. Dong, *ibid.* **127** (1980), 2722.
69. J. K. Srivastava and J. B. Wagner, Jr., *ibid.* **131** (1984), 196C, Abstract No. 463RNP.
70. B. Agius, S. Rigo, F. Rochet, M. Froment, C. Maillot, H. Roulet, and G. Dufour, *Appl. Phys. Lett.* **44** (1984), 48.
71. A. G. Revesz, *J. Electrochem. Soc.* **126** (1979), 122.
72. W. D. Mackintosh and H. H. Plattner, *ibid.* **124** (1977), 396.
73. P. J. Jorgensen, *J. Chem. Phys.* **37** (1962), 874; *J. Electrochem. Soc.* **114** (1967), 820.
74. D. N. Modlin and W. A. Tiller, *J. Electrochem. Soc.* **132** (1985), 1659.
75. D. O. Raleigh, *ibid.* **113** (1966), 782; **115** (1968), 111.
76. S. J. Laverty and W. D. Ryan, *Int. J. Electronics* **26** (1969), 519.
77. T. G. Mills and F. A. Kroger, *J. Electrochem. Soc.* **120** (1973), 1582.
78. J. R. Ligenza and W. G. Spitzer, *J. Phys. Chem. Solids* **14** (1960), 131.
79. E. W. Sucov, *J. Am. Ceram. Soc.* **46** (1963), 14.
80. R. Haul and G. Dümbgen, *Z. Elektrochem.* **66** (1962), 636.
81. E. L. Williams, *J. Am. Ceram. Soc.* **48** (1965), 190.
82. K. Muehlenbachs and H. A. Schaeffer, *Can. Mineralogist* **15** (1977), 179.
83. H. A. Schaeffer, private communication.
84. F. Rochet, B. Agius, and S. Rigo, *J. Electrochem. Soc.* **131** (1984), 914.
85. J. C. Mikkelsen, Jr., *Appl. Phys. Lett.* **39** (1981), 601, 903; **40** (1982), 336.
86. J. C. Mikkelsen, Jr., *J. Electron. Mater.* **11** (1982), 541.
87. R. Pfeffer and M. Ohring, *J. Appl. Phys.* **52** (1981), 777.
88. D. R. Wolters, *J. Electrochem. Soc.* **127** (1980), 2072.
89. Y. Z. Lu and Y. C. Cheng, *J. Appl. Phys.* **56** (1984), 1608.
90. D. L. Griscom, in *Defects and Their Structure in Nonmetallic Solids*, B. Henderson and A. E. Hughes, Eds. Plenum Press, New York, 1976, p. 323.
91. N. F. Mott, *Adv. Phys.* **26** (1977), 363.
92. G. Lucovsky, *Phil. Mag.* **B39** (1979), 513.
93. M. J. Dignam, in *Comprehensive Treatise of Electrochemistry*, Vol. 4, J. O'M. Bockris, B. E. Conway, E. Yeager, and R. E. White, Eds., Plenum Press, New York, 1981, p. 247.
94. P. J. Caplan, E. H. Poindexter, B. E. Deal, and R. R. Razouk, *J. Appl. Phys.* **50** (1979), 5847.
95. E. H. Poindexter, P. J. Caplan, B. E. Deal, and R. R. Razouk, *ibid.* **52** (1981), 879.
96. F. P. Fehlner, *J. Electrochem. Soc.* **131** (1984), 1645.
97. N. Jakowski and H. Glaefeke, *Thin Solid Films* **36** (1976), 195.
98. P. F. Schmidt and W. Michel, *J. Electrochem. Soc.* **104** (1957), 230.
99. M. J. Verkerk et al., *J. Electrochem. Soc.* **130** (1983), 70, 78.
100. P. J. Jorgensen, *J. Chem. Phys.* **49** (1968), 1594.
101. G. C. Jain, A. Prasad, and B. C. Chakravarty, *J. Electrochem. Soc.* **126** (1979), 89.
102. J. R. Ligenza, *J. Appl. Phys.* **36** (1965), 2703.
103. A. K. Ray, *Thin Solid Films* **84** (1981), 389.
104. T. Sugano, *ibid.* **92** (1982), 19.
105. D. L. Pulfrey, F. G. M. Hathorn, and L. Young, *J. Electrochem. Soc.* **120** (1973), 1529.

106. J. Siejka and J. Perriere, in *Plasma Synthesis and Etching of Electronic Materials*, R. P. Chang and B. Abeles, Eds. (Mater. Res. Soc. Symp., **38**) Materials Research Society, Pittsburgh, PA, 1985, p. 427.
107. S-i. Kimura, E. Murakami, K. Miyake, T. Warabisako, H. Sunami, and T. Tokuyama, *J. Electrochem. Soc.* **132** (1985), 1460.
108. E. D. Atanassova et al., *Thin Solid Films* **100** (1983), 131; *Solid-State Elec.* **25** (1982), 781.
109. M. Watanabe and A. Tooi, *Japan. J. Appl. Phys.* **5** (1966), 737.
110. J. Dylewski and M. C. Joshi, *Thin Solid Films* **35** (1976), 327.
111. K. I. Kirov, E. D. Atanasova, S. P. Alexandrova, B. G. Amov, and A. E. Djakov, *ibid.* **48** (1978), 187.
112. S. W. Chiang, Y. S. Liu, and R. F. Reihl, *Appl. Phys. Lett.* **39** (1981), 752.
113. Y. S. Liu, S. W. Chiang, and F. Bacon, *ibid.* **38** (1981), 1005.
114. R. Oren and S. K. Ghandhi, *J. Appl. Phys.* **42** (1971), 752.
115. S. A. Schafer and S. A. Lyon, *J. Vac. Sci. Technol.* **21** (1982), 422.
116. E. M. Young and W. A. Tiller, *Appl. Phys. Lett.* **42** (1983), 63.
117. I. W. Boyd, J. I. B. Wilson, and J. L. West, *Thin Solid Films* **83** (1981), L173.
118. I. W. Boyd and J. I. B. Wilson, *Appl. Phys. Lett.* **41** (1982), 162.
119. M. M. Atalla, E. Tannenbaum, and E. J. Scheibner, *Bell Sys. Tech. J.* **38** (1959), 749.
120. E. Kooi, *J. Electrochem. Soc.* **111** (1964), 1383.
121. E. Kooi, *Philips Res. Repts.* **20** (1965), 306, 578, 595; **21** (1966), 477.
122. E. Kooi, *The Surface Properties of Oxidized Silicon*, Springer-Verlag, New York, 1967.
123. B. E. Deal, *J. Electrochem. Soc.* **127** (1980), 979.
124. H. P. Vyas, G. D. Kirchner, and S. J. Lee, *ibid.* **129** (1982), 1757.
125. A. G. Revesz, *ibid.* **124** (1977), 1811.
126. D. L. Griscom, M. Stapelbroek, and E. J. Friebele, *J. Chem. Phys.* **78** (1983), 1638.
127. N. Mott, M. Pepper, S. Pollitt, R. H. Wallis, and C. J. Adkins, *Proc. Roy. Soc.* (*London*) **A345** (1975), 169.
128. G. H. Frischat, *Ionic Diffusion in Oxide Glasses*, Trans Tech Publications, Bay Village, OH, 1975.
129. D. W. Hess and B. E. Deal, *J. Electrochem. Soc.* **124** (1977), 735.
130. J. Monkowski, R. E. Tressler, and J. Stach, *ibid.* **125** (1978), 1867; **126** (1979), 1129.
131. K. Ehara, K. Sakuma, and K. Ohwada, *ibid.* **126** (1979), 2249.
132. W. A. Weyl, *Colored Glasses*, Society of Glass Technology, Sheffield, 1951, Reprinted by Dawson's of Pall Mall, London, 1959, p. 98ff.
133. C. Hashimoto, S. Muramoto, N. Shiono, and D. Nakajima, *J. Electrochem. Soc.* **127** (1980), 129.
134. J. Steinberg, *ibid.* **129** (1982), 1778.
135. S. Iwamatsu, *ibid.* **129** (1982), 224.
136. S. S. Cohen, *ibid.* **130** (1983), 929.
137. A. G. Revesz, *Thin Solid Films* **41** (1977), L43.
138. Part V of *Insulating Films on Semiconductors*, M. Schulz and G. Pensl, Eds., Springer-Verlag, Berlin, 1981, pp. 150–179.
139. H. Lefevre and M. Schulz, *Appl. Phys.* **12** (1977), 45.
140. A. Chantre, M. Kechouane, and D. Bois, in *Laser and Electron Beam Interactions with Solids 1981*, B. R. Appleton and G. K. Celler, Eds., North-Holland, Amsterdam, 1982, p. 325.
141. E. H. Snow and B. E. Deal, *J. Electrochem. Soc.* **113** (1966), 263.
142. E. H. Snow and M. E. Dumesnil, *J. Appl. Phys.* **37** (1966), 2123.
143. N. M. Johnson, D. K. Biegelsen, and M. D. Moyer, *J. Vac. Sci. Technol.* **19** (1981), 390.

144. R. R. Razouk and B. E. Deal, *J. Electrochem. Soc.* **126** (1979), 1573.
145. P. O. Hahn, *J. Vac. Sci. Technol.* **A2** (1984), 574.
146. Y. Hayafuji and K. Kajiwara, *J. Electrochem. Soc.* **129** (1982), 2102.
147. S. S. Wong, C. G. Sodini, T. W. Ekstedt, H. R. Grinolds, K. H. Jackson, S. H. Kwan, and W. G. Oldham, *ibid.* **130** (1983), 1139.
148. C. P. Ho and J. D. Plummer, *ibid.* **126** (1979), 1516, 1523.
149. A. S. Grove, *Physics and Technology of Semiconductor Devices,* John Wiley & Sons, New York, 1967.
150. L. I. Maissel and R. Glang, *Handbook of Thin Film Technology*, McGraw-Hill Book Co., New York, 1970, p. 11–44.
151. W. A. Pliskin, in *Physical Measurement and Analysis of Thin Films*, E. M. Murt and W. G. Guldner, Eds., Plenum Press, New York, 1969, p. 168.
152. F. L. Galeener, J. C. Mikkelsen, Jr., R. H. Geils, and W. J. Mosby, *Appl. Phys, Lett.* **32** (1978), 34.
153. J. C. Mikkelsen, Jr. and F. L. Galeener, *ibid.* **37** (1980), 712.
154. J. C. Mikkelsen, Jr., F. L. Galeener, and W. J. Mosby, *J. Electron. Mater.* **10** (1981), 631.

10

CLOSING SUMMARY

It is apparent from the preceding chapters that low-temperature oxidation is a complex and fascinating phenomenon. The role of vitreous oxides in controlling the kinetics of reaction is considered to be critical in this process.

The reaction of oxygen with a clean metal surface has been conveniently divided into three parts: chemisorption and oxygen incorporation, island growth during a transition period, and slow thickening of the three-dimensional oxide.

This last part of the overall process depends upon the structure of the oxide, whether it is a network former, intermediate, or modifier. A generalized comparison of these three classifications is given in Table 10.1. The properties compared are self-consistent within each class of oxide, which strengthens their relevance to oxidation at low temperatures. For instance, the movement of anions and oxygen pressure dependence both occur in network formers which have open structures.

Application of this structural distinction to oxidation in general depends upon experimental conditions. Cleanliness, temperature, pressure, and time all affect the stability of a vitreous oxide. The best example of this is found in silicon oxidation. Here, structure appears to allow both molecules and anionic species to be involved in the transport process.

At high temperatures and pressures, oxygen solubility in SiO_2 is sufficient to support both the initial linear kinetics limited by interfacial reaction and the subsequent diffusion-limited parabolic kinetics. Oxygen incorporation is at the Si/SiO_2 interface. When pressure is reduced, a field is applied, or temperatures are low, then the molecular model no longer appears to apply. Instead, an electric field, either spontaneous or applied, enhances movement of a charged oxygen defect. Kinetics then follow the inverse logarithmic law.

It is one purpose of the present work to point out that it would be very

TABLE 10.1. GENERALIZED COMPARISON OF OXIDE TYPES

Type of oxide	Major type of bonding	Structure at 25°C	Chief self-diffusing ion	Oxidation kinetics	Major interface dependence	Role of water
Network former	Covalent	Vitreous network	Anions	Inverse logarithmic	Oxygen pressure	Weakens network by breaking covalent bonds
Intermediate	Mixed	Microcrystalline	Anions and cations	Inverse logarithmic	Pressure and crystallography	Mixed effect
Modifier	Ionic	Polycrystalline	Cations	Direct logarithmic	Crystallographic orientation of metal	Strengthens ionic lattice by hydrogen bonding

fruitful to extend the success of vitreous oxide in solving problems of silicon technology to other high-temperature systems. Extension to protective coatings, as for tantalum alloys [1], can give dramatic impetus to control of metallic corrosion as pointed out by Hoar [2].

REFERENCES

1. C. M. Packer and R. A. Perkins, *J. Less-Common Met.* **37** (1974), 361.
2. T. P. Hoar, *J. Electrochem. Soc.* **117** (1970), 17C.

INDEX